Ukrainian Armour and Fighting Vehicles

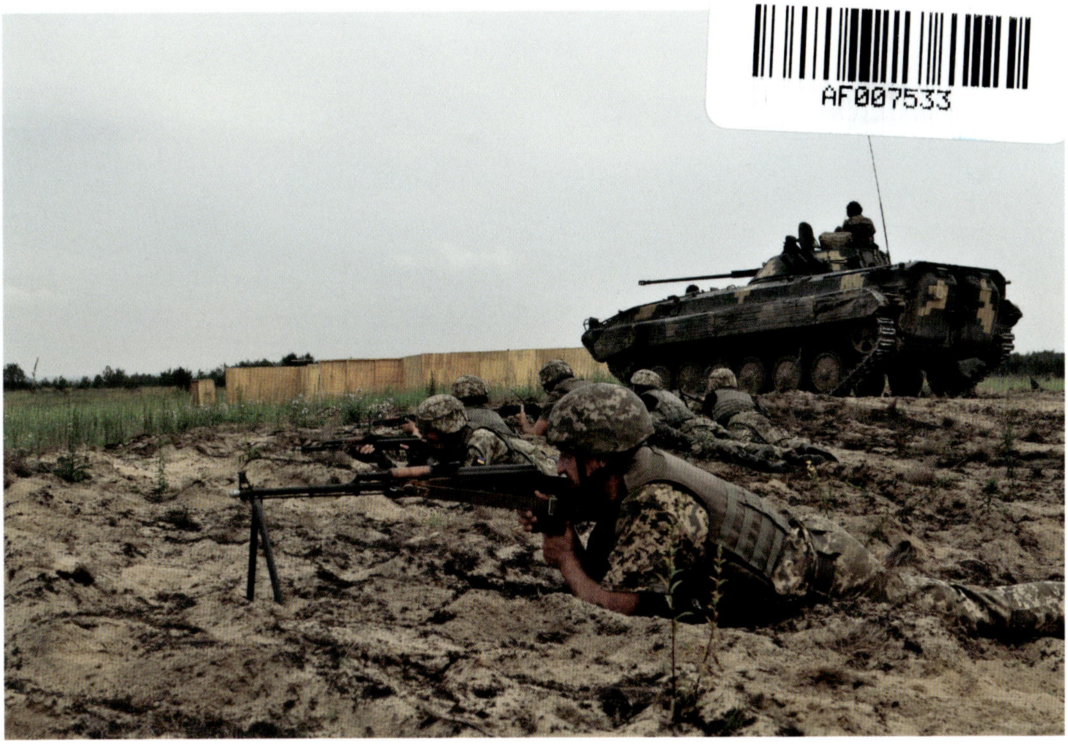

Craig Allen

KEY Books

MILITARY VEHICLES AND ARTILLERY SERIES, VOLUME 7

Title page image: A BMP-2 provides support to Ukrainian infantry during a platoon live-firing. (Capt. Scott Kuhn, public domain, via Wikimedia Commons)

Contents page image: A battery of BM-21s in firing position. The Ukrainian military often use individual Grads in 'shoot and scoot' missions. (Vitaly V. Kuzmin, CC BY-SA 4.0 https://creativecommons.org/licenses/by-sa/4.0, via Wikimedia Commons)

Back cover image: Ukrainian troops move up to the front in BMP-2s. (Ministry of Defense of Ukraine, CC BY-SA 2.0 https://creativecommons.org/licenses/ by-sa/2.0, via Wikimedia Commons)

Published by Key Books
An imprint of Key Publishing Ltd
PO Box 100
Stamford
Lincs PE9 1XQ

www.keypublishing.com

The right of Craig Allen to be identified as the author of this book has been asserted in accordance with the Copyright, Designs and Patents Act 1988 Sections 77 and 78.

Copyright © Craig Allen, 2023

ISBN 978 1 80282 596 1

All rights reserved. Reproduction in whole or in part in any form whatsoever or by any means is strictly prohibited without the prior permission of the Publisher.

Typeset by SJmagic DESIGN SERVICES, India.

Contents

Introduction 4

Chapter 1 Ukrainian Equipment 5
The T-64 5
The T80U and T84 Oplot 7
The BTR-4 Bucephalus IFV 9
Varta Light APC 13

Chapter 2 Soviet Equipment 19
The T-72 19
The 2S1 122mm SPG 21
BMP 1/2 IFV 23
BM-21 Grad 25

Chapter 3 UK Equipment 29
The Challenger 2 29
AS90 SPG 34
The M270 MLRS 36
The Mastiff 38
Husky TSV 41
The Snatch Land Rover 42

Chapter 4 US Equipment 46
Abrams M1 46
The Bradley M2 IFV 49
M109 SPG 51
M142 HIMARS 53

Stryker ICV 55
M113 APC 58
The Humvee 61

Chapter 5 German Equipment 63
Leopard 1 63
Leopard 2 65
Gepard Anti-Aircraft System 68
PzH 2000 SPG 70
The Marder IFV 71

Chapter 6 French Equipment 74
AMX 10 RC 74
Ceaser SPG 76
VAB APC 78

Chapter 7 Polish Equipment 80
PT-91 Twardy 80
Krab SPG.. 82

Chapter 8 Other Nations 84
The M-55S Tank 84
Bushmaster PMV 86
ACSV 88
Senator APC 89
Swedish CV90 IFV 91

Introduction

In the ongoing conflict between Russia and Ukraine, the two protagonists have relied heavily on their armoured units. For both sides, this has largely meant employing ex-Soviet tanks and armoured fighting vehicles (AFVs) alongside a few more recent types. Like the Russians, the Ukrainians have their own tank production facilities, which in the past have manufactured designs such as the T-64 from the Soviet era and more modern tanks such as the T-84 Oplot. Furthermore, with large numbers of Russian-produced vehicles already in use within the country, the Ukrainians have found little problem in repairing and repurposing tanks and military equipment abandoned on the battlefield. Then there are the huge quantities of vehicles and artillery supplied by the Western countries. Initially, this excluded modern main battle tanks and the armour was mostly donated by former Warsaw Pact (1955) countries such as Poland and the Czech Republic. This mostly came in the form of modernised T-72s, a tank the Ukrainians are familiar with and well able to use and maintain. As the war has progressed, however, their inventory has become more diverse, with an eclectic mix of ex-Soviet and Western types, which now includes tanks such as the Challenger 2 and Leopard 2. In this work, I will attempt outline the tanks, self-propelled artillery and armoured vehicles currently in use. With new equipment continually being added to the list, it is difficult to keep up, but I have tried to cover all the major types in use at the time of writing.

A Ukrainian reconnaissance/drone team pictured with their Snatch Land Rover. (Craig Allen)

Chapter 1
Ukrainian Equipment

The T-64

This tank came about through the rivalry between the Soviet Union's two design teams at Uralvagonzavod in Russia and Kharkiv in Ukraine. The rival teams were seeking to replace the older generation T-55s and T-62s with their own design of tank. Chief designer Leonid Kartsev at Uralvagonzavod took a progressive approach that eventually led to the T-72. Meanwhile, Alexander Morozov in Kharkiv went for a more radical design featuring several advanced features that became the T-64. The concept behind Morozov's advanced design was to make the tank as compact as possible, thereby lowering its silhouette. It featured a completely new hull design, while an autoloader for the 125mm gun allowed for a smaller turret and reduced the crew numbers from four to three. The Soviet military chiefs were initially impressed with the new tank, however, trials quickly showed problems with both its V-45 engine and autoloader. Despite these difficulties, the tank was eventually approved for series production.

The T-64 went through several developments during the 1960s and 1970s including improvements to the engine, drive-train and fire control system. The final BM versions were fitted with Explosive Reactive Armour (ERA) and were powered by an improved 6TD engine. The Kharkiv tank team also mounted the 2A461M-1 125mm gun, which was capable of firing Kobra radio-guided missiles. From around 1999, the Ukrainian military began to modernise its older T-64BMs to create the T-64BM Bulat. This upgraded version featured improved reactive armour, the 125mm KBA3 gun and a new night sight, all produced in Ukraine. Power was derived from the 850hp 5TDFM multi-fuel diesel engine capable of speeds of up to 43mph. Production of the new model was slow, however, with less than 100 available by the start of the war in the Donbas in 2014. This means many of the T-64s currently in use are still of the older models, although in recent years two further variants have emerged. The T-64BM2 features a new 1,000hp 6TD-1 engine, thermal sight and improved fire control system. The T-64BV model adds improved communications equipment, GPS navigation and Nizh reactive armour modules. In 2019, these upgraded models began to emerge from the Malyshev Factory in Lviv. The T-64, with its advanced and compact design, would eventually become the basis for the more modern T-80 and T84 tanks.

Specifications	
Model	T-64BM Bulat, (Upgraded T-64)
Manufacture	Malyshev Factory
Country	Ukraine
Year	2010
Engine	1,000hp 6TD-1 engine
Fuel	Diesel
Range	500km
Suspension	Torsion bar
Top Speed	60kph
Armament	125mm KBA3 main gun, 12.7mm and 7.62mm machine guns (MGs)
Capacity	3 crew
Weight	45 tons

A Ukrainian T-64BV. This is the latest model of the tank, featuring thermal imaging, digital communications and new Nizh reactive armour. (Mil.gov.ua, CC BY 4.0 https://creativecommons.org/licenses/by/4.0, via Wikimedia Commons)

A T-64BV in frontline use. Note the side-mounted Explosive Reactive Armour (ERA) tiles and foliage camouflage. Sometimes individual T-64s would take on the Russian tank columns. (Vitalii Strutynskyi, CC BY-SA 2.0, https://creativecommons.org/licenses/by-sa/2.0, via Wikimedia Commons)

A T-64BV on the move in the Donbas. Note the infrared searchlight to the left of the main gun and cutting devices on the front wings. (Ministry of Defense of Ukraine, CC BY-SA 2.0 https://creativecommons.org/licenses/by-sa/2.0, via Wikimedia Commons)

The T80U and T84 Oplot

The Russian-designed T-80 was essentially a modernised version of the T-64. It was fitted with a gas turbine engine and entered service during the 1980s. In parallel with the production of the T-80, the Ukrainian Morozov design team developed a more reliable diesel-powered version, the T-80U. This featured full-length side skirts, Kontakt-5 reactive armour and a 125mm smoothbore gun. Produced at the Kharkiv plant, some 300 examples of the tank came into the possession of the Ukrainian military upon the break up of the Soviet Union. Following an upgrade of welded turrets, 1,200hp 6TD-2 diesel engines and a Ukrainian-produced 125mm main gun plus reactive armour, these tanks received the designation T-84. As well as going into service with the Ukrainian Armed Forces, these tanks were also offered for export sale. The T84U was an upgraded version fitted with armoured side skirts, integrated reactive armour, thermal imaging sight, GPS and a laser range finder. An improved version that came about in the 1990s, the T-80UM, featured a thermal imaging sight and later the 2A46M-4 version of the 125mm gun. The latest version, the T-84 Oplot, first produced in 2001, is essentially a T-84U with a new design of turret with separate crew and ammunition compartments plus a bustle-mounted autoloader. While significant numbers of older T-84s are available to the Ukrainian Armed Forces, only a small number of T84 Oplots have been produced so far. This means that T-80Us have been pressed into current service to make up tank numbers. Several T-80s have also been captured from the Russians, and Ukrainian forces have refurbished these tanks for their own use.

Specifications	
Model	T-84 Oplot
Manufacture	Malyshev Factory
Country	Ukraine
Year	2001
Engine	1,200hp 6TD-2
Fuel	Diesel
Range	450km
Suspension	Torsion bar
Top Speed	70kph
Armament	125mm KBA3 main gun, 12.7mm & 7.62mm MGs
Countermeasures	Shtora-1
Capacity	3 crew
Weight	48 tons

A T-84U of the Ukrainian 3rd Tank Brigade. These are essentially diesel-powered variants of the T-80 and provided the basis for the more modern T-84 Oplot. (Mil.gov.ua, CC BY 4.0 https://creativecommons.org/licenses/by/4.0, via Wikimedia Commons)

T-84 on the move. These tanks were upgraded by the Malyshev Factory with 1,200hp diesel engines and Kontakt-5 reactive armour. (7th Army Training Command from Grafenwöhr, Germany, public domain, via Wikimedia Commons)

A T-84 Oplot on display in Kyiv. These tanks are a modernised version of the T-84U, with a new turret and thermal imaging, GPS and integrated reactive armour. They have only been produced in small numbers. (VoidWanderer, CC BY-SA 4.0 https://creativecommons.org/licenses/by-sa/4.0, via Wikimedia Commons)

A T-84U concealed in woods. Significantly outnumbered by Russian armour, the Ukrainian Armed Forces have to rely on stealth and superior tactics. (Ministry of Defense of Ukraine, CC BY-SA 2.0 https://creativecommons.org/licenses/by-sa/2.0, via Wikimedia Commons)

The BTR-4 Bucephalus IFV

The BTR-4 is a product of the Morozov Machine Building Design Bureau based in Kharkiv and was initially developed as a private venture in the early 2000s. The vehicle is essentially a further evolution of the Soviet-era BTR series of eight-wheeled armoured personnel carriers (APCs). It was redesigned to eliminate some of the BTR's deficiencies, such as the rear-mounted engine that hinders troop access. The BTR-4 uses a mid-engine layout, making it more adaptable and features Western influence in its design. The prototype was first unveiled at the Aviasvit military exhibition held in Ukraine in 2006 and was initially intended for export, with several countries showing interest, including Iraq and Kazakhstan. The Ukrainian military was already using the older generation BTR-80 and modernised BTR-94, and there was no extra funding available at the time for the purchase of BTR-4s. Produced by state defence manufacturer UkrOboronProm, the first orders went to Iraq designated as the BTR-4E (E for export). However, the Iraqi military was not completely happy with the vehicle due to quality control issues and cancelled the order after some 80 examples had been delivered. The remainder of the order went to the Ukrainian Armed Forces, which eventually found the necessary funds, with the model being officially adopted in 2012. Kazakhstan also found a number of faults with the original design, and the order were never fulfilled. In wake of the Russian assault in the Donbas in 2014, the Ukrainian military ordered some 150 BTR-4s. Production would continue right up to the second Russian invasion of 2022, with the vehicle seeing wide-scale use with Ukrainian frontline units.

Although externally similar in appearance to the Soviet BTR series of 8-wheeled APCs, the BTR-4 uses a different mid-engine layout with the crew compartment situated at the rear. The hull is divided into three main compartments – cab, engine and troop – each separated by an armoured bulkhead. The crew consists of a driver commander and gunner, while between seven and nine fully equipped troops can be carried depending on the armament and configuration. Troop access is via the two rear doors and roof hatches, while side doors are provided for the crew. The vehicle is powered by a Ukrainian-built 3TD diesel engine, reaching 500bhp and driving through a five-speed auto transmission. The armour only offers protection from small-arms and shell fragments, however, the frontal plates can withstand strikes from 12.7mm (.50 calibre) rounds. Add-on armour panels are available, and the vehicle has underbelly protection against mines. Cage armour can often be seen fitted to operational vehicles and this offers a degree of protection from Russian anti-tank missiles. A fire suppression system is fitted, along with full nuclear, biological and chemical (NBC) protection for the crew. The vehicle has a good power-to-weight ratio and mobility is excellent. This is enhanced by a central inflation system allowing tyre pressures to be altered when crossing soft or boggy ground. For amphibious use, the BTR-4 is powered by two water jets, and a bow plate is deployed before entering the water. The vehicle has a maximum 110kph road speed while the jets propel it at up to 10kph in water. Armament consists of the Grom or Parus remote weapons stations fitted with a 30mm cannon, grenade launcher and 7.62mm MG. These turrets can also be fitted with up to four Baryer anti-tank guided missile (ATGM) launchers. Further options include the Shkval mount with a similar armament but only two ATGM launchers and the BAU 232, which features a pair of 23mm auto-cannons. However, most BTR-4s in service appear to be equipped with the BM-7 Parus turret mounted with a pair of ATGMs. The addition of these remote weapon stations put the BTR-4 in the category of a wheeled infantry fighting vehicle (IFV) rather than just a basic armoured personnel carrier. Several variants of the BTR-4 have been produced, including specialist reconnaissance, fire support, command, ambulance and recovery versions.

Specifications	
Model	BTR-4
Manufacturer	UkrOboronProm
Country	Ukraine
Year	Officially adopted 2012
Engine	3TD diesel engine or DEUTZ EURO III
Transmission	5-speed automatic
Fuel	Diesel
Range	690km
Suspension	8x8 wheel independent
Top Speed	110kph (road), 10kph (water)
Armament	30mm cannon, 30mm grenade launcher, 7.62mm MG plus up to 4x Baryer ATGM
Armour	Welded steel plus modular add-on armour
Capacity	3 crew plus 7–9 dismounts
Weight	18–24 tons

A BTR-4 pictured at a Ukrainian military display. Note the prominent bow plate and 30mm remote weapons station. (Artemis Dread, CC BY-SA 4.0 https://creativecommons.org/licenses/by-sa/4.0, via Wikimedia Commons)

Right: The cab and driver's position of the BTR-4. Note the automatic-gear selector. (VoidWanderer, CC BY-SA 4.0 https://creativecommons.org/licenses/by-sa/4.0, via Wikimedia Commons)

Below: This view shows the troop doors and one of the rear-mounted propellers for amphibious use. (VoidWanderer, CC BY-SA 4.0 https://creativecommons.org/licenses/by-sa/4.0, via Wikimedia Commons)

Later versions of the BTR-4, like this Varen model, feature a rear ramp replacing the troop doors. (VoidWanderer, CC BY-SA 4.0 https://creativecommons.org/licenses/by-sa/4.0, via Wikimedia Commons)

This image of a Russian BTR-80 clearly shows the lineage of the BTR-4. Note the top-mounted troop hatches. (Vitaly V. Kuzmin, CC BY-SA 4.0 https://creativecommons.org/licenses/by-sa/4.0, via Wikimedia Commons)

Varta Light APC

The Ukrainian-produced Varta is a mine-proofed light armoured personnel carrier (APC) or Mine-Resistant Ambush Protected (MRAP) in the same mould as the British-made General Dynamics Foxhound. Designed and built by Ukrainian Armor Company, the vehicle is equipped with a classic V-shaped hull to deflect blast. The 560 steel armoured body can withstand strikes from small arms and shell fragments, and the vehicle is powered by a six-cylinder turbo-diesel engine delivering 300hp and driving through an eight-speed manual transmission. With its 4x4 system, the Varta has excellent cross-country mobility and features run-flat tyres with a central inflation system. The crew are provided with blast attenuating seating, and there is room for up to eight dismounts in the rear. Access is via a pair of armoured doors on each side or the large rear troop door, and there are roof mounted hatches. Both a gunners protection kit (GPK) and armoured turret are available to mount either a 7.62mm or 12.7mm MG for local protection. The small, armoured side windows are also equipped with firing ports, enabling embarked infantrymen to use their personal weapons. The Varta has also been seen mounting the Javelin Missile System. The vehicle comes equipped with front and rear cameras, an automatic fire suppression system and air conditioning as standard. The front-mounted winch has a nine-ton capacity and can be used as an aid to recovery. Designed to meet NATO standards, the Varta came into service with Ukrainian Forces in 2018. The very similar Novator produced by the same company is a crew cab variant featuring a rear-load area. Both of these light APCs are useful for delivering drone and reconnaissance teams and as general troop transports.

A Varta armoured personnel carrier (APC) on display in Kyiv. Note the GPK turret and front-mounted winch; the cables attached to the wings are designed to deflect undergrowth in close country. (VoidWanderer, CC BY-SA 4.0 https://creativecommons.org/licenses/by-sa/4.0, via Wikimedia Commons)

Specifications

Model	Varta APC
Manufacturer	Ukraine Armor
Country	Ukraine
Year	Officially adopted 2018
Engine	6-cylinder turbo-charged 300hp
Transmission	8-speed manual
Fuel	Diesel
Range	600km
Suspension	4x4 fully independent
Top Speed	100kph
Armament	7.62mm or 12.7 MG, Javelin ATGW
Armour	Welded 560 steel
Capacity	2 crew plus 8 dismounts
Weight	17.2 tons

This view illustrates the rear-mounted step for troop access and the prominent side windows with firing ports. (ukrarmor.com)

The Novator crew-cab variant features a rear load area with nylon canopy but is otherwise mechanically very similar to the Varta APC. (ukrarmor.com)

Under New Management

Making use of the enemy's armoured vehicles and equipment is nothing new. What is perhaps new is the sheer scale of abandoned heavy weapons left in the wake of Russian retreats in the current conflict. The initial withdrawal from Kyiv brought a rich bounty of T72 tanks, BTR APCs and GAZ trucks having either run out of fuel or simply broken down by the roadside. The Ukrainian farmers became unexpected allies in recovering much of this bounty, and videos of tractors hauling away abandoned Russian armour became common on YouTube. In the advance in the Kharkiv region, even the very latest T90 tanks and 2S19M2 152mm SPGs have been discovered in perfectly serviceable condition by the victorious Ukrainian troops. Images of abandoned Russian positions, poorly constructed and strewn with rubbish reminded me of my own experience in Iraq, a combat indicator of poorly trained and disciplined troops. In the later advances from Kharkiv, the so-called 're-deployment' of Russian Forces was actually more akin to a rout, with large quantities of weaponry and ammunition left behind. It now appears that elements of the 1st Guards Tank Army were also present, but these so-called elite troops fled as readily as the rest. Some of this hardware was in good condition and ready for immediate use, and any tanks, vehicles and guns too damaged to salvage were cannibalised for spare parts. In Izium alone, ten T80 tanks and several 152mm 2S3 Howitzers were captured along with a complete Tor-M1 Air Defense System. The welcome donation of all this armour and equipment would be of limited benefit without the ability

A Russian BM-21 GMRL was captured by the Ukrainian Armed Forces in the Donbas. (Mil.gov.ua, CC BY 4.0 https://creativecommons.org/licenses/by/4.0, via Wikimedia Commons)

to actually use it. In this, the Ukrainians have an ace up their sleeve, as not only are they familiar with this ex-Soviet equipment but are also able to maintain it. The country possesses its own tank factories where all these captured AFVs can be readily repaired and upgraded for further use. For example, it now appears the armoured spearhead that broke through the front in the East included numerous ex-Russian T72-B3M tanks under new management. One disadvantage of this mix of ex-Soviet and Western-supplied equipment is the difficulty of supplying multiple ammunition types. However, helpfully the Russians also left behind large quantities of 122mm and 152mm shells, which the Ukrainian military desperately needed.

So much equipment has been captured that the Russian Armed Forces have actually become Ukraine's largest supplier of heavy weapons. Independent intelligence consultants Oryx have recorded the capture of 460 Russian main battle tanks, 92 SPGs, 448 infantry fighting vehicles, along with numerous other pieces of equipment. Their assessments rely solely on video evidence, so the real figures are likely considerably higher. While some items are retained by frontline units for their own use, the Ukrainian military has also become proficient at backloading this captured weaponry into the supply chain. There are reports of Ukrainian infantry units such as the Carpathian Battalion becoming fully mechanized simply from captured equipment. Other units have boosted their artillery component with Russian Howitzers and SPGs. This rich haul combined with Western-supplied systems is helping to power the Ukrainian advance and neutralize the Russian advantage in artillery. Turning all this captured hardware back on the enemy is also undoubtedly helping morale.

Scenes of jubilant Ukrainian troops atop captured Russian tanks have become an enduring symbol of success in this war. It would be hard to find a conflict where enemy weaponry has been acquired on such a scale by the opposing side.

Ukraine had been running low on ammunition for its Soviet-era artillery, but are now amply re-plenished from captured Russian stocks. (Mil.gov.ua, CC BY 4.0 https://creativecommons.org/licenses/by/4.0, via Wikimedia Commons)

Above: This captured T62M is evidence of older equipment being pulled out of storage by the Russians due to the heavy losses in tanks. (Mil. gov.ua, CC BY 4.0 https://creativecommons.org/licenses/by/4.0, via Wikimedia Commons)

Right: A modern Russian 2S19M2 152.4mm SPG was captured during the Kharkiv counteroffensive. (Mil. gov.ua, CC BY 4.0 https://creativecommons.org/licenses/by/4.0, via Wikimedia Commons)

A soldier from the 17th Brigade paints the white cross of Ukraine on a captured T-72B3 tank. (Mil.gov.ua, CC BY 4.0 https://creativecommons.org/licenses/by/4.0, via Wikimedia Commons)

Early in the war, images of Ukrainian farmers towing away abandoned Russian armour became a common sight. (Eurasiantimes.com)

Chapter 2

Soviet Equipment

The T-72

The T-72 was originally conceived by the Kartsev design team as a stop-gap model during the Cold War but has gone on to become one of the most widely used main battle tanks of the modern era. Many former Warsaw Pact countries were left significant numbers of these tanks as a legacy after the collapse of the Soviet Union, and the T-72 has been widely exported and still remains in demand by those nations unable to afford the latest Western designs. The Ukrainian military inherited over 1,000 T-72s, however, a large number of these were quickly sold off to Third World Countries. This was due to the decision to concentrate on the T-64 produced in Ukraine-based factories. The remaining 300 to 400 examples were placed in storage, however tank losses incurred in the Donbas fighting from 2014 meant they began to be reactivated. In 2018, the Kiev Armaments Plant started refurbishing these stored T-72s to T-72AMT standard. The T-72AMT incorporates many of the upgrades found on the later T-64 models, including an improved engine, reactive and side skirt armour, thermal imaging sights and modern communications equipment. So far, around 40–50 have been delivered to Russian units, however over 100 T-72s have been captured by the Ukrainian forces in the recent fighting. Many of these are of the latest T-72B3 model and are often abandoned in good condition. The T-72B3 has some of the upgrades of the T-80 and T-90 tanks, such as Kontact-5 reactive armour. It also features a new engine, gunner's sight and improved VHF communications. Meanwhile, former Warsaw Pact countries such as Poland and the Czech Republic have begun supplying legacy T-72s from their own stocks. These are mostly T-72M1s, the export version of the tank, and many have already received some upgrades. For example, the Polish government has sent PT-91 Twardy tanks, which are T-72M1s fitted with new engines, fire control and communications equipment. The Ukrainian military has plenty of experience in operating the T-72 and have set up mobile workshops close to frontlines to repair and re-use enemy tanks that fall into its hands.

Specifications	
Model	T-72B3
Manufacturer	Uralvagonzavod
Country	Russia
Year	2013
Engine	V-92S2F 1,130hp
Fuel	Diesel
Range	500km
Transmission	Hydraulically assisted; 7 forward and 1 reverse gear
Suspension	Torsion bar
Top Speed	70kph
Armament	125mm smoothbore gun
Capacity	3 crew
Weight	45 tons

The T-72 has been steadily upgraded over the years. This B3 model in Ukrainian service is fitted with ERA and may be a captured example. (Vitaly V. Kuzmin, CC BY-SA 4.0 https://creativecommons.org/licenses/by-sa/4.0, via Wikimedia Commons)

The T-72B3 was updated with improved sights, a new autoloader and more powerful engine. It was also fitted with Kontakt-5 ERA tiles. (www.unian.ua)

A pair of Ukrainian T-72B1s operating together. Note the additional ERA tiles and slat armour for extra protection from ATGWs. (Mil.gov.ua, CC BY 4.0 https://creativecommons.org/licenses/by/4.0, via Wikimedia Commons)

The 2S1 122mm SPG

The 2S1 is the self-propelled version of the 122mm D-30 Howitzer and first entered service with the Soviet Armed Forces in the early 1970s. It is based on the tracked MT-LB chassis and the main armament is mounted in a welded steel turret. The 2S1 uses the same 122mm ammunition types as the D-30, including HE-Frag, cluster and smoke rounds. Loading is semi-automatic using a powered rammer, and the 2S1 is capable of a maximum rate of fire of four to five rounds a minute. An optical PG-2 sight used for indirect fire and accuracy is generally good, with the weapon having an effective range of 15.2km. The 2S1 is powered by a YaMZ-238V 240hp diesel engine and the vehicle is amphibious with some preparation. The 2S1 has seen action in several conflicts including the Russo-Afghan and Iran-Iraq wars. Significant numbers of these veteran self-propelled guns (SPGs) remain in use by both the Ukrainian and Russian armed forces. However, they are now long in the tooth and lack the range and accuracy of modern Western designs.

Specifications	
Model	2S1 SPG
Manufacturer	Soviet Union
Country	Russia
Year	1971
Armament	2A-18 122mm Howitzer
Protection	Welded steel armour
Engine	YaMZ-238V 240hp
Fuel	Diesel
Range	500km
Transmission	Manual
Suspension	Torsion bar
Top Speed	60kph
Capacity	4 crew
Weight	16 tons

A 2S1 in Ukrainian service. These veteran machines are now older than most of their crews. (Mil.gov.ua, CC BY 4.0 https://creativecommons.org/licenses/by/4.0, via Wikimedia Commons)

A Ukrainian 2S1 caught at the moment of firing. Accuracy is reasonably good, but it can't match modern Western designs with digital fire control. (Mil.gov.ua, CC BY 4.0 https://creativecommons.org/licenses/by/4.0, via Wikimedia Commons)

A 2S1 in action somewhere in Ukraine. The 122 main gun can use a full range of ammunition including HE-Frag, cluster and smoke rounds. (Mil.gov.ua, CC BY 4.0 https://creativecommons.org/licenses/by/4.0, via Wikimedia Commons)

BMP 1/2 IFV

The BMP-1 was the original IFV designed and built in the Soviet Union during the Cold War era. Produced in large numbers, it was widely exported to the Warsaw Pact countries and their allies. The tracked chassis features a front-mounted engine, while the rear crew compartment is accessed via two armoured doors and four roof hatches. Unusually, the rear doors also house the vehicles auxiliary fuel tanks. The turret mounts a 73mm Grom main gun plus a launcher for the AT-3A Sagger ATGM. The tracked hull is constructed of steel armour, and the vehicle's low silhouette means the interior is cramped for the eight infantry dismounts. Power comes from a six-cylinder 300hp diesel engine driving through a five-speed manual gearbox. Fully amphibious, the BMP-1 has good cross-country mobility and can reach speeds of up to 65kph on road. The vehicle's side armour offers protection from small arms of 7.62mm calibre, while the slopped frontal armour can withstand hits from 12.7mm rounds. The BMP-1 has seen action in several conflicts, beginning with the Yom Kippur War (6–25 October 1973). It was employed by the Soviets in Chechnya and Afghanistan in the 1980s and 1990s, where it proved vulnerable to rocket-propelled grenade (RPG) strikes and improvised explosive devices (IEDs). During the first Gulf War (2 August 1990–28 February 1991), the Americans found it could be successfully penetrated by 25mm AP rounds from their Bradley IFVs.

The lessons learned in combat and the increasing obsolescence of the 73mm main gun led to the development of the BMP-2, which went into service in the 1980s. This featured a new dual-feed 30mm auto-cannon with coaxial 7.62mm PK MG. Some later versions also featured an ASG-30mm grenade launcher and Kornet anti-tank missile system. The internal layout was slightly changed, with a two-man turret and space for only seven infantrymen. Protection was also improved so the vehicle could withstand up to 12.7 calibre from the side and 25mm hits on the sloped frontal armour. The increased weight was compensated by a more powerful engine, and mobility is similar to the earlier BMP-1. The Ukrainian military was already equipped with legacy BMP-1s, some of which it has updated to BMP-1 UMD configuration. Many more have been captured from the Russians and large numbers of BMP-1s and 2s have been supplied by Poland, the Czech Republic and Germany. Despite this, combat losses have left the Ukrainian military short of IFVs, which is why a request has been made for modern Western types.

Specifications	
Model	BMP-1
Manufacturer	Kurganmashzavod
Country	Russia
Year	1966
Armament	73mm Grom plus AT3 Sagger ATGW
Protection	Welded steel armour
Engine	UTD-20, 6-cylinder 300hp
Fuel	Diesel
Range	600km
Transmission	Manual 4-speed
Suspension	Torsion bar with hydraulic shocks
Top Speed	65kph
Capacity	3 crew
Weight	13.3 tons

The BMP-1 was one of the first modern Infantry Fighting Vehicles (IFVs) and was widely exported. This example was photographed in Iraq. (Lance Corporal Lanham, public domain, via Wikimedia Commons)

A Ukrainian modernised BMP-1 UMD with a new turret mounting a 30mm auto-cannon and equipped with a modern fire control system. (VoidWanderer, CC BY-SA 4.0 https://creativecommons.org/licenses/by-sa/4.0, via Wikimedia Commons)

A BMP-2 in service in Ukraine. Note the low silhouette and cramped turret. (Staff Sgt. Adriana Diaz-Brown, public domain, via Wikimedia Commons)

BM-21 Grad

The BM-21 is the successor to 'Stalin Organ' – a slang military term for a multiple rocket launcher – of World War Two fame and is a truck-mounted multiple-launch rocket system. Developed in the 1960s, it features a bank of 40 122mm rocket tubes mated to a Ural 6x6 truck. The launch tubes are rifled and high-explosive fragmentation, incendiary and chemical warheads are available. The latest rocket types include cluster munitions and have a range of some 30km. The accuracy is poor, however, so the Grad should be considered as an area weapon. The BM-21 is operated by a three-man crew, and a telescopic sight and collimator is used for sighting purposes. The system can be brought into action within three minutes and fired electrically from the cab. The BM-21 has been widely exported and has seen action in numerous wars including Vietnam, Afghanistan, and Chechnya. It is employed by both sides in the current Russo-Ukrainian conflict, although the Ukrainians are now receiving far more accurate Western rocket artillery.

Specifications	
Model	BM-21 Grad
Manufacture	Soviet Union
Country	Russia
Year	1963
Engine	IL-375 180hp
Transmission	5-speed manual, all-wheel drive
Fuel	Diesel
Range	750km
Suspension	Leaf springs
Top Speed	75kph
Armament	40 x 122mm rocket launchers
Capacity	3 crew
Weight	13.7 tons

A BM-21 Grad in Ukrainian service with the launch tubes in the transit position. Note the hand-operated elevation drum. (Ministry of Defense of Ukraine, CC BY-SA 2.0 https://creativecommons.org/licenses/by-sa/2.0, via Wikimedia Commons)

The Grad uses a collimating sighting system but should be considered as an area weapon due to its lack of accuracy. (VoidWanderer, CC BY-SA 4.0 https://creativecommons.org/licenses/by-sa/4.0, via Wikimedia Commons)

A Grad unleashes its rockets! The BM-21 has a range of some 30km and warheads include high-explosive fragmentation, incendiary and cluster. (sputniknews.lat)

The Artillery Duel – Russian D-30 Gun vs US M777 Howitzer

The 122mm D-30 first entered service with Soviet forces in the 1960s and still remains in use as the standard towed artillery piece of Russian forces. Large numbers are also employed by the Ukrainian forces, although acquiring an ammunition supply for these guns has become increasingly difficult. In order to aid the Ukrainian side, the US government has supplied large numbers of the 155mm M777 Howitzer, a far more modern gun that fires standard NATO ammunition. So how do the two artillery pieces measure up against each other in actual use?

The D-30 may be an older weapon, but it has some interesting features, including a three-legged gun mount and barrel-mounted tow link. It features a semi-automatic sliding wedge breach and uses a hydropneumatic recoil system. The ammunition uses separate loading charges and rocket assisted projectiles are also available. The long .35-calibre barrel features a prominent muzzle brake and the effective range is 15.3km (21km with assisted munitions). A well-trained seven-man crew can achieve firing rates of 6–7rpm. However, due to the excessive muzzle blast, a lanyard must be used, and normal rates are closer to 5–6rpm. The D-30 is a relatively simple design that is easy to maintain in the field without special tools. Despite this, there have been reports of some Russian guns suffering shattered barrels due to extreme wear. The Ukrainians are also wearing out the barrels of their D-30s and have difficulty in sourcing replacements.

The US M777 155mm gun is actually a British design originally developed by Vickers as a light Howitzer in the 1980s. It was then further refined for American use by BAE Land Systems and came into service as the M777 in 2005. The extensive use of titanium in its construction serves to

The ex-Soviet 122mm D-30 Howitzer is used by both sides and is a robust weapon that can be maintained in the field. (rznonline.ru)

keep the weapon light. The unconventional gun carriage uses a system of stabilisers and ground spades, and like the D-30 features a barrel-mounted towing eye. Also, like the D-30, the M777 has a low profile when mounted. The long .39-calibre barrel features an unusual design of rectangular muzzle brake with prominent baffles and the weapon features a digital fire control system. The effective range with standard NATO ammunition types is 21km, which can be extended to 40km using special Excalibur nose-bleed munitions. The gun uses an eight-man crew, which can maintain a maximum rate of fire of 4rpm, although 2rpm is more usual. Ukrainian gunners have been very impressed with the M777, although barrel wear caused by excessive levels of firing can affect accuracy. Changing barrels can't be done in the field, and maintenance is likely to become an ongoing problem.

The US M777 offers greater range and accuracy, while the Russian D-30 has a better rate of fire and is more robust. Both of these weapons are suffering excessive wear and tear given the current levels of combat. Keeping them in action and supplying them with sufficient ammunition will therefore be a challenge to both sides in this conflict.

The M777 in action. This modern gun has helped the Ukrainians redress the imbalance in artillery systems. (Mil.gov.ua, CC BY 4.0 https://creativecommons.org/licenses/by/4.0, via Wikimedia Commons)

Chapter 3

UK Equipment

The Challenger 2

The updated version of the Challenger 1 began as an in-house venture by Vickers Armstrong in the mid-1980s. The Ministry of Defence (MoD) soon became interested in this development and ordered a prototype for testing. This led to an initial order for 140 of the new model, with production beginning in 1993 and the Challenger 2 entering service in 1998. Eventually, a little over 400 of the new version were built for the British Army, replacing the Challenger 1. The new model was an extensive update of the original, although outwardly similar in appearance. It was now armed with the L30A1 version of the 120mm rifled gun in a redesigned turret and was fitted with an improved fire control system. Second-generation Chobham armour was fitted for an all-up weight of some 75 tons. Despite an increase in weight, it had a similar cross-country performance to Challenger 1 thanks to its improved hydro-pneumatic suspension. Challenger 2s were employed on operational service in Bosnia and Kosovo and saw action in the invasion of Iraq in 2003. The Challenger 2 went to war with the 7th Armoured Brigade, and the new tanks made short work of any Iraqi T-55s they encountered. Post-conflict, the tanks were upgraded with an add-on passive armour package and the Challenger 2 remains one of the best-protected Main Battle Tanks (MBTs) currently in service. At the time of writing, 14 tanks, sufficient for a squadron, have been delivered to Ukraine. The crews are currently undergoing training in the UK as of March 2023, and the Challenger 2 should prove more than a match for the Russian counter-armour. The one caveat is the limited numbers that are available, although the UK's defence minister has stated he is open to the idea of supplying more Challenger 2s.

A Challenger 2 undergoes sight testing before a range practice in the UK. Note the side skirts and turret-mounted grenade dischargers. (Craig Allen)

Specifications	
Model	Challenger 2
Manufacturer	BAE Land Systems
Country	UK
Year	1998
Engine	Perkins CV12-6A V12
Fuel	Diesel
Range	550km
Transmission	TN54E epicyclic, 6 forward and 2 reverse gears
Suspension	Hydro-pneumatic
Top Speed	59kph
Armament	L30A1 120mm rifled gun, 7.62 chain-gun coax, 7.62 general-purpose machine gun (GPMG)
Capacity	4 crew
Weight	75 tons

The 120mm rifled gun combined with take-off gross weight (TOG) gives the Challenger 2 impressive hitting power and accuracy. (Craig Allen)

A Challenger 2 in Iraq in 2003. It performed well during the war-fighting phase, and no Challengers were penetrated by enemy fire. (Craig Allen)

The Challenger 2 remains very capable and is one of the best protected NATO tanks. The large drum at the rear is an external fuel tank. (Craig Allen)

CVRT

The Combat Vehicle Reconnaissance Tracked (CVRT) has been around since the 1970s and is still in service with British Armed Forces in updated form. In addition to the Scimitar, which is armed with a 30mm Rarden cannon the CVRT is configured as an APC (Spartan), armoured ambulance (Samaritan) and recovery vehicle (Samson). There is also a command version (Sultan) with a distinctive box body with raised roofline. The CVRT was one of the first AFVs to feature aluminium armour to save weight. Originally equipped with a Jaguar 4.2-litre petrol engine, many have now been retrofitted with a Cummins 5.9 diesel. This is mated to an auto-gearbox and gives the vehicle increased range. Many of the CVRTs used in Iraq and Afghanistan also received additional protection including anti-RPG slat/bar armour. The UK has supplied over 100 CVRTs to Ukraine whilst others have been purchased privately. These are mainly Spartan, Samaritan and Samson types. Fast and with good cross-country mobility, they have proved useful to the Ukrainian military, especially the Spartans, which have been employed to deploy drone and reconnaissance teams.

Specifications	
Model	CVRT (Spartan)
Manufacturer	Alvis/Bea Systems
Country	UK
Year	1970 to present
Protection	Aluminium armour
Engine	Cummins BTA 190hp
Fuel	Diesel
Transmission	Auto
Suspension	Torsion bar
Range	450km
Top Speed	80kph
Armament	Crew weapons
Capacity	3 crew plus 4 dismounts
Weight	9 tons

A Spartan armoured patrol car (APC) during driver training. Fast and manoeuvrable, these veteran armoured vehicles have proved popular in Ukraine. (Airwolfhound from Hertfordshire, UK, CC BY-SA 2.0 https://creativecommons.org/licenses/by-sa/2.0, via Wikimedia Commons)

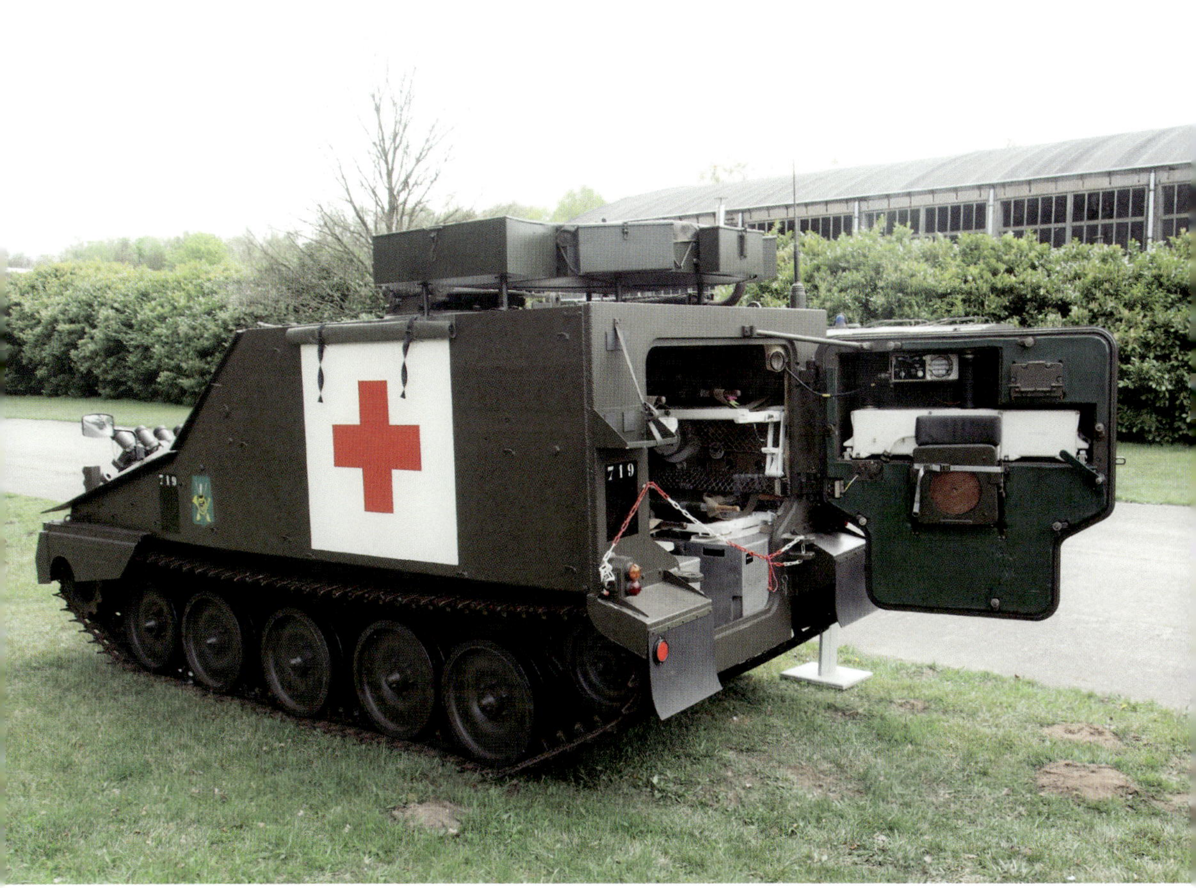

The Samaritan ambulance version uses the high box body of the Sultan command vehicle and can carry up to four stretcher cases. (Alf van Beem, public domain, via Wikimedia Commons)

British troops evacuate a casualty to a Samaritan during training. They have proved equally useful in this role in Ukraine. (Craig Allen)

AS90 SPG

The AS90 is the British Army's standard 155mm self-propelled Howitzer and has been in service since the early 1990s. Designed and built by Vickers, it replaced the Abbot 105mm and US-built M109 155mm SPG. The 155mm L31 gun has an effective range of 24km, although this can be extended to 30km with assisted munitions. Gun laying is achieved via the automatic gun laying system (AGLS), while upgrades include a digital ballistic computer and BAE Systems laser inertial artillery pointing system (LINAPS). Automatic loading enables rapid firing, and three rounds can delivered within ten seconds in burst mode. The welded steel hull is proof against small arms and shell splinters, and the tracked chassis gives the AS90 good cross-country mobility. The vehicle is powered by the Cummins VTA-903 660hp turbo-charged diesel engine mated to an automatic gearbox. With its powerful engine and hydro-pneumatic suspension, it can reach speeds of up to 53kph. The 155mm gun can be brought into action in less than a minute, and the barrel can be clamped from within the vehicle to allow quick changes of position. In January 2023, the Ukrainian military was promised 30 AS90s, which will be a significant boost to its artillery inventory.

Specifications	
Model	AS90 SPG
Manufacturer	Vickers
Country	UK
Year	1992
Armament	155mm L31 .39-calibre gun
Protection	Welded steel armour
Engine	Cummins VTA-903 660hp
Fuel	Diesel
Range	420km
Transmission	Automatic
Suspension	Hydropneumatic
Top Speed	53kph
Capacity	5 crew
Weight	45.7 tons

AS90s on the move on a training area in the UK. Note the substantial barrel clamp and the impressive size of the vehicle. (Richard Watt/MOD, OGL v1.0OGL v1.0, via Wikimedia Commons)

An AS90 during firing practice in the UK. Note the rear hatch used for loading additional ammunition. (Cpl Timothy Jones/MOD, OGL v1.0OGL v1.0, via Wikimedia Commons)

This shot of a crewmember bringing up ammunition illustrates the high angles of elevation possible for the AS90. (Sgt Si Longworth RLC (Phot)/MOD, OGL v1.0OGL v1.0, via Wikimedia Commons)

The M270 MLRS

The M270 GMLRS is a 227mm American-designed (guided) multiple launch rocket system ([G]MLRS) manned by the Royal Artillery. The vehicle features twin rocket pods and has a range of some 30km. Each disposable pod contains six rockets, which can be fired individually or as a single 'fire for effect' of 12 rockets. There is no launching rail, and the rockets are fired directly from their pods. Reloading is power-assisted and can be accomplished in 8–10 minutes by a trained crew. The rockets are fitted with a guidance system using a combination of inertial navigation and jam-resistant GPS and have a range of up to 80km. Accuracy is reportedly extremely high, with the near-vertical terminal trajectory. The M31A1 GMLRS rocket is fitted with a single blast warhead and the fuse has three modes; point, delay and air-burst. As of March 2023, the UK has donated a small number of M270s along with stocks of missiles. The French and German governments have also supplied Ukrainian forces with MLRS and these have proved extremely useful in allowing the Ukrainian military to target the Russian's vulnerable rear defence areas.

Specifications	
Model	M270B1 MLRS (UK version)
Manufacturer	Lockheed Martin
Country	UK
Year	2003
Engine	Cumins 500hp
Fuel	Diesel
Transmission	Auto
Suspension	Torsion bar
Range	485km
Top Speed	39mph
Armament	M269 Launcher Loader Module
Capacity	3 crew
Weight	24.5 tons

The M270 multiple launch rocket system (MLRS) has been a game changer for the Ukrainian military. Its 80km range bringing the Russian rear defences within reach. (censor net)

UK Equipment

Right: An MLRS is brought into firing position in Ukraine. The rockets can be fired individually or in salvos and are GPS-guided for extreme accuracy. (Олексій Мазепа / АрміяInform, CC BY 4.0 https://creativecommons.org/licenses/by/4.0, via Wikimedia Commons)

Below: An M270 pictured on range at Otterburn, UK. Note the armoured shutters have been closed over for firing. (Cpl Jamie Peters RLC/MOD, OGL v1.0OGL v1.0, via Wikimedia Commons)

Ukrainian soldiers train with the M270 MLRS in the UK prior to the transfer of the weapons systems. (Militarnyi)

The Mastiff

The Mastiff is the British version of the US Cougar mine-protected vehicle (MPV) and was first fielded in Afghanistan late in 2006. In UK service, it featured a Bowman radio fit and an appliqué armour package. Able to carry eight fully equipped infantrymen in addition to its two-man crew and mount a selection of weaponry for local defence, it was an immediate success. This quickly led to additional purchases in 2007 and was followed by the development of an enhanced version, the Mastiff 2. This incorporated several improvements to the design, including the use of dyneema unidirectional (UD) armour. Stronger axles and suspension were fitted to cope with the tough Afghan terrain. The turret now featured powered traverse for the gunner and fire suppression systems were fitted to the engine and fuel tanks. The driver wasn't forgotten and gained a new thermal imager to aid situational awareness. Although the same v-shaped hull was employed along with blast attenuating seating, overall protection was improved in the new version. As well as the troop carriers, Mastiffs were employed by the Royal Engineer IED teams and a special mine clearance version was developed. They were also used as command vehicles and armoured ambulances, and the Royal Air Force (RAF) utilised the Praetorian, as it was known in RAF service, a version fitted with a camera system mounted on a telescopic mast. There was also a dedicated logistic version, the Wolfhound, which was used mainly for convoy work and as an artillery tractor for the 105 Light Gun. Like many of the other specialist types developed for Afghanistan, the vehicles were put up for foreign sales in 2017, and recently a number have been gifted to Ukraine as part of the military support package.

UK Equipment

Specifications	
Model	Mastiff 2
Manufacturer	Force Protection (NP Aerospace, Coventry, UK, for modifications)
Country	UK
Year	2009
Engine	Caterpillar C7 330hp
Fuel	Diesel
Range	966km
Transmission	Allison 3500SP Automatic
Suspension	6x6 wheeled
Top Speed	55mph
Armament	7.62 GPMG, .5 HMG, 40mm H&K GMG
Capacity	2 plus 8 dismounts
Weight	28 tons

A Mastiff in use with the Ukrainian Marines. Note the slat armour and extra protection for the gunner's position. (АрміяInform, CC BY 4.0 https://creativecommons.org/licenses/by/4.0, via Wikimedia Commons)

The Mastiff was originally procured for use in Afghanistan. Note the anti-rocket-propelled grenade (RPG) slat armour and numerous antennae for the electronic countermeasures (ECM) equipment. (Craig Allen)

Access to the troop compartment is via the twin armoured doors at the rear. Note the rear-mounted camera in its armoured housing above the doors. (Craig Allen)

The Wolfhound is the logistic version of the vehicle and features a rear load area for transporting stores and equipment. (Andrew Linnett/MOD, OGL v1.0OGL v1.0, via Wikimedia Commons)

Husky TSV

Manufactured by Navistar Defence, the Husky is a medium-armoured high-mobility tactical support vehicle (TSV) based on the International MXT model. The vehicle was designed specifically for the British Army for deployment in Afghanistan in 2009. It is a four-wheel drive medium platform TSV designed for maximum mobility and protection. The vehicle can accommodate a four-man crew, including the driver and commander. The Husky was designed to support patrols operating in lower threat areas and was produced in three main variants: utility, ambulance and command post vehicles, as well as a heavy recovery version. The utility variant is equipped with a flatbed, while the command and ambulance variations have enclosed cabs at the rear. An ambulance developed with enhanced protection entered service in 2010, along with the command post vehicle. The vehicle features a GPMG for local protection while a remote weapon station can be fitted capable of mounting a 12.7mm heavy machine gun (HMG). The armoured hull is designed for blast protection and can be up-armoured with either an A or B-type appliqué armour kit. The Husky is powered by MaxxForce D6.0L V8, which provides 340bhp and drives through an Allison auto-transmission. There is a central tyre inflation system operated from the cab and the vehicle has a maximum road speed of 70mph. The Husky is another Afghan veteran that will prove useful to the Ukrainians. It is capable of taking on a number of battlefield support roles and can tow lighter artillery pieces.

Specifications	
Model	Husky TSV
Manufacturer	Navistar Defence
Country	UK
Year	2010
Engine	MaxxForce D6.0L V8
Fuel	Diesel
Transmission	Allison 2500 SP automatic
Suspension	4x4
Top Speed	70mph
Range	400 miles
Armament	7.62 GPMG or 12.7 HMG
Capacity	4 crew
Weight	6.8 tons

The Husky was specifically designed as a tactical support and patrol vehicle for use in Afghanistan but is finding new roles in Ukraine. (Andrew Linnett/MOD, OGL v1.0OGL v1.0, via Wikimedia Commons)

Husky TSV have been supplied to Ukraine as part of the UK's military assistance package along with other Afghan veterans such as the Mastiff and Wolfhound. (Andrew Linnett/MOD, OGL v1.0OGL v1.0, via Wikimedia Commons)

The Snatch Land Rover

The Snatch Land Rover was originally designed for use as a patrol vehicle in Northern Ireland to replace the earlier ad hoc types with bolt-on protection kits. It was based on the heavy-duty 110 Wolf chassis and developed with the aid of the Ricardo engineering company. The composite armoured body offered only limited protection from blast, and the vehicles was never intended to withstand powerful IEDs. This led to the Snatch gaining a bad reputation with the media after losses in Iraq and Afghanistan. As a result, these Land Rovers were subsequently up-armoured and the original V8 petrol engines replaced by 300tdi diesels. They were eventually replaced by purpose-built designs such as Foxhound and, as a result, many have been released as surplus. With a two-man crew, the Snatch can carry up to four dismounts with a roof-mounted hatch for top cover. Armament relies on the troop's personal weapons and the Snatch is useful for patrolling in low-threat environments. Ukraine has purchased a number of these Land Rovers for use as armoured ambulances, and they have also been used to evacuate civilians from the frontline areas.

Specifications	
Model	Snatch (CAV 100)
Manufacturer	Land Rover/Ricardo
Country	UK
Year	2006
Engine	3.9 V8 or 300tdi
Fuel	Petrol/Diesel
Transmission	5-speed manual
Suspension	4x4
Top Speed	60mph
Range	320 miles
Armament	Personal weapons
Capacity	2 crew plus 4 dismounts
Weight	3.9 tons

The Snatch Land Rover was upgraded for use in Iraq and Afghanistan, with additional armour and 300tdi diesel engines. This model is also fitted with frangible wheels similar to the WMIK Land Rover. (Cpl Russ Nolan RLC, OGL v1.0OGL v1.0, via Wikimedia Commons)

Large numbers of Snatch Land Rovers like this one have been released as surplus and are now being re-purposed for use by the Ukrainian Armed Forces. (tanks-a-lot.co.uk)

Tank Buster
Why the 'Next Generation Light Anti-Tank Weapon' (NLAW) has proved such a game-changer in the war in Ukraine

One of the enduring images of the conflict in Ukraine will be the sight of devastated Russian armour, destroyed by shoulder-launched anti-armour systems. The US has given Javelins and the Germans Panzerfaust 3s to the Ukrainian military, but it is the UK NLAW that has made the most impact on the battlefield. Loved by the Ukrainian troops and praised by their officers, the NLAW has cut a swathe through Russian armoured columns. But why has it proved so effective? After all, man-portable anti-tank weapons are nothing new. The German Panzerfaust was the bane of allied armour in World War Two, and Russia's own RPG is the most widely used weapon of the type. So, what is the difference with this new breed of anti-armour weapon?

First, some history, as the NLAW was preceded by a long line of anti-tank systems used by British Armed Forces. In the 1980s, the British relied on the 66mm LAW backed up by the effective but cumbersome 84mm Carl Gustaff. Meanwhile, specialist anti-tank units were provided with the excellent wire-guided MILAN missile system. While MILAN remained effective, the older weapons were soon outclassed by modern developments in tank armour. This led to the development of the LAW 80, whose 94mm warhead could penetrate some 700mm of conventional rolled armour. I used this weapon during my own service days and was frankly less than impressed. An ergonomic disaster and with a sighting system that was difficult to use, I wouldn't have liked to rely on one with a T72 bearing down on me. Its use in Iraq highlighted further issues, plus the weapon could be defeated by modern reactive armour. The search soon began for a more effective replacement, and in 2002 a joint UK/Swedish venture led to the weapon system that would become the NLAW. Developed by the Swedish Saab Bofors Dynamics and the UK's Thales Air Defence in conjunction

The NLAW has proved a game changer in Ukraine, with poor Russian tactics giving tank-hunting teams a field day. (Aljazeera)

with the Ministry of Defence, it was designed to replace the older generation of shoulder-launched weapons and give the infantry a tank-killing capability.

The NLAW uses 'predicted line of sight', which effectively means it's a 'fire and forget weapon', the operator able to seek cover after launching the missile. Capable of either direct or top attack modes, it is effective from 20m out to a maximum of 1,000m. The missile is launched using a low-powered ignition system, the main rocket motor only initiating several metres from the launcher. This means it can be fired successfully from buildings or confined spaces such as trench systems. In top attack mode, the missile is detonated by a proximity fuse launching the 150mm HEAT warhead down onto the least well-protected area of the tank. The effectiveness of all this can be seen in the many images of destroyed and blasted Russian armour.

We know NLAW is an effective tank-buster, but this still doesn't explain the extent of Russian losses. Other battlefield factors are also behind Ukrainian success with the weapon, not least a seeming lack of tactical awareness from Russian troops (see 'Armourgeddon' section). Pushing columns of bunched-up armour down limited axis and an over-reliance on roads has created a target-rich environment. Armour operating without infantry support or top cover that might detect and take on the Ukrainian tank-hunting teams is another obvious mistake. Then there is the failure to disperse during halts and entering built-up areas without proceeding infantry. These are all basic tactical errors from a modern professional army that exposed Russian military ineffectiveness. Further losses inflicted by armed drones and artillery, plus mechanical breakdown and fuel shortages have also left some Russian units ineffective. Perhaps the real step change is NLAWs ability to completely destroy a modern MBT placing a new level of capability into the hands of the infantry.

Chapter 4
US Equipment

Abrams M1

In the 1970s, both the US and Germany were working on new tank designs that could defeat the latest generation of Soviet armour. They had initially collaborated on a common design, but the partnership ended with the Germans going on to develop the Leopard 2. The Americans, meanwhile, produced the MBT-70, which featured a number of new technologies including hydro-pneumatic suspension. This design proved something of a dead end, however, and by the mid-1970s General Motors and Chrysler were asked to submit new proposals under the designation XM1. The Chrysler version of the XM1 was eventually chosen, partly for political reasons. This was surprising given it featured a fuel-hungry gas turbine engine rather than a more efficient diesel power plant. The main armament chosen was a licence-built version of the Royal Ordnance 105mm L7 gun, and the design was the first to feature Chobham composite armour. The XM1 was re-designated as the Abrams M1 and officially entered service with the US Army in 1980. From 1986, the Rheinmetall 120mm L/44 smoothbore gun was adopted for the M1A1 version of the tank. The Abrams has been steadily upgraded over the years, including fitting depleted uranium armour and improved fire control and thermal viewing equipment. The tank saw extensive operational service in the two Gulf Wars, where it proved more than a match for Iraqi T-72s. The Abrams is heavily armoured, and although a number were disabled by shoulder-launched missiles in Iraq, none were penetrated. The 120mm gun is capable of knocking out enemy tanks at ranges in excess of 2,500m. With an all-up weight of over 60 tons, it is one of the heaviest main battle tanks, which presents its own problems with transport and bridge crossing.

The Abrams M1 has been thoroughly battle tested. This one seen during operations in Iraq: note the very large turret and raised rear hull. (Joseph A. Lambach, U.S. Marine Corps, public domain, via Wikimedia Commons)

In January of 2023, the US announced it would be sending 31 Abrams M1s to Ukraine as part of a military assistance package. These are to be the Abrams M1A2 version of the tank, which feature the latest optical sights and fire control systems. They are also a digitally enabled platform allowing the rapid transfer of data. However, the depleted uranium armour will be removed from these tanks before delivery, as the technology with which it is built is still considered a secret. How effective the Abrams actually proves to be remains to be seen. It will potentially take until 2024 before they can be delivered, and the crews trained to operate them. Perhaps their real value is in giving Germany the political cover it required to release its own Leopard 2s, which will be far more useful to the Ukrainians (see 'Leopard 2' section).

Specifications	
Model	Abrams M1A2
Manufacturer	Detroit Tank Plant
Country	United States
Year	1990
Armament	20mm XM256 smoothbore gun
Protection	Chobham armour
Engine	1500hp gas turbine engine
Fuel	JP-8 fuel (kerosene)
Range	426km
Transmission	Allison DDA X-1100-3B auto
Suspension	Torsion bar
Top Speed	42kph
Capacity	4 crew
Weight	64 tons

This frontal view illustrates the width of the turret in relation to the hull. Note the front-mounted recovery bars and twin hatches for the commander and gunner. (Staff Sgt Aaron Allmon, public domain, via Wikimedia Commons)

Despite its weight, the Abrams has good mobility, but the gas turbine engine is extremely fuel hungry and the M1 is more complex to maintain than other NATO tanks. (Chad Menegay, public domain, via Wikimedia Commons)

This rear view of the M1 shows the raised rear hull and side skirt armour. Note the large stowage basket on the rear of the turret. (Craig Allen)

The Bradley M2 IFV

The Bradley is the US Army's primary IFV and first came into service in the early 1980s. Equipped with a powerful 25mm chain gun and TOW launcher, it is able to give fire support to its accompanying infantry and take on enemy armour. The standard M2 Bradley carries a three-man crew plus six fully equipped infantrymen. The M3 version is designed for scouting and carries additional TOW missiles with more radio equipment and fewer dismounts. Large numbers of both the Infantry and Cavalry/Scout versions of the Bradley were employed in the two Gulf Wars. They generally performed well in action, although proved vulnerable to RPG strikes and IEDs. An upgrade programme was therefore initiated based on combat experience. This included fitting the Bradleys with reactive armour plus improved navigation and communications equipment, as well as other improvements. These improved versions are designated M2A3, and in further upgrades have received digital command-and-control systems, energy-absorbing seating and an improved target acquisition system for the TOW launcher. The M3 version has also received similar upgrades. Despite various initiatives to replace the Bradley, it remains in service with the US Armed Forces and Saudi Arabia. In January 2023, it was announced that 109 Bradley M2A3s are to be provided to Ukraine.

Specifications	
Model	Bradley M2A3
Manufacturer	BAE Systems Land
Country	United States
Year	1981
Armament	25mm M242 Bushmaster chain gun plus 2x BGM-71 TOW
Protection	Spaced laminate armour plus add-on ERA panels
Engine	Cummins VTA-903T 600hp
Fuel	Diesel
Range	400km
Transmission	HMPT-500 hydro-mechanical
Suspension	Torsion bar
Top Speed	56kph
Capacity	3 crew plus 6 dismounts
Weight	27.6 tons

The Bradley M2 was extensively combat-tested in the Gulf Wars and offers a good balance of mobility, protection and firepower. (Sgt. 1st Class Johancharles Van Boers, public domain, via Wikimedia Commons)

This image of Bradleys on a road move gives a good view of the driver's and commander's positions. Note the forward recovery points and side skirt armour. (7th Army Joint Multinational Training Command from Grafenwöhr, Germany. (CC BY 2.0 https://creativecommons.org/licenses/by/2.0, via Wikimedia Commons)

Troops rapidly debus from the Bradley's rear loading ramp. The standard M2 can carry six dismounts in addition to its three-man crew. (Sgt. Richard Jones, public domain, via Wikimedia Commons)

M109 SPG

Production of the veteran M109 self-propelled Howitzer began in 1962, and it first saw active service in the Vietnam War. Exported to several allied nations, it was used by the Israelis in the Yon Kippur War in 1973 and by the British in the first Gulf War. The vehicle's main armament is the 155mm M126 L23 gun with a range of 14.6km. Powered by a Detroit Diesel 8V71T 450hp engine, it can reach speeds of up to 65kph on the road and has an amphibious capability. The six-man crew includes two loaders, and a rapid rate of fire of up to four rounds per minute can be maintained for short periods. The M109 has been steadily upgraded over the years, and the A2 version saw the adoption of a .39-calibre M189 gun with a longer barrel and greater range. The M109A6 Paladin that entered service in the 1990s featured a more advanced fire control system in addition to modern digital communications. The latest M109A7 version has received further improvements, including an autoloader and upgraded fire control system. The M109 is now an old weapon system but remains in service with numerous armies and is still considered effective. Ukraine has acquired updated versions of the M109 from various sources and initially purchased a batch of ex-Belgian machines from a private company. The UK has also supplied a batch of 20 refurbished former Belgium Army M109s, while Norway has provided 22 of its modernised M109A3GN vehicles. These weapons are already in use by the Ukrainians and will help redress the Russian's considerable advantage in artillery systems.

Specifications	
Model	M109A2
Manufacturer	United Defence LP
Country	US
Year	1979
Armament	155mm .39-calibre M189 gun
Protection	Welded steel armour
Engine	Detroit 8V71T 450hp
Fuel	Diesel
Range	420km
Transmission	Allison; 4 forward 2 reverse
Suspension	Torsion bar
Top Speed	65kph
Capacity	6 crew
Weight	27.5 tons

An M109 pictured at a display in Ukraine. These older generation self-propelled guns (SPGs) are still in widespread use and many have been updated with modern fire control systems. (Stephencdickson, CC BY-SA 4.0 https://creativecommons.org/licenses/by-sa/4.0, via Wikimedia Commons)

Above: An M109 SPG in action in Ukraine. Firing standard NATO 155mm ammunition, and with a range of up to 14km, these are still an effective weapon system. (Mil.gov.ua, CC BY 4.0 https://creativecommons.org/licenses/by/4.0, via Wikimedia Commons)

Left: With its tracked chassis, the M109 has good cross-country mobility and can cope with the Ukrainian mud. Note the long barrel, which is locked down in transit mode. (armyinform.com.ua)

An M109 opens fire. This is a German version, and the vehicle has been widely exported and is still in service around the world. (Uwe from bei Hof, Franken, CC BY-SA 2.0 https://creativecommons.org/licenses/by-sa/2.0, via Wikimedia Commons)

M142 HIMARS

The M142 High Mobility Artillery Rocket System (HIMARS) is a truck-based missile system developed in the 1990s for the US military. It features a single MLRS rocket pod mounted on a 6x6 FMTV chassis with armoured cab. The M142 can fire either 227mm GMLRS rockets or the longer-range Army Tactical Missile System (ATACMS). The fire control system and electronics are interchangeable with those of the tracked M270 MLRS and the crew training is the same. Rapid reloading is possible via a built-in crane system, and the rockets come as a six-round pod. The HIMARS was developed to be readily transported by air and has good cross-country mobility with its wheeled chassis. Initial US deliveries of HIMARS to Ukraine amounted to 16 vehicles with a further 18 promised. The US has provided the Ukrainians with M30A1 missiles, which feature an improved warhead but have yet to approve the ATACMS, which has a reach of up to 300km. It is hard to understate the importance of these weapons to Ukraine, as their accuracy and long range brings its enemy's rear areas within reach.

Specifications	
Model	M142 HIMARS
Manufacture	Lockheed Martin
Country	US
Year	2010
Engine	Caterpillar 3116 ATAAC 290hp
Transmission	Allison 3700 SP 7-speed auto
Fuel	Diesel
Range	480km
Suspension	Parabolic leaf springs
Top Speed	85kph
Armament	1 x MLRS pod (6 x launchers)
Capacity	3 crew
Weight	45 ton

A column of M142 High Mobility Artillery Rocket Systems (HIMARS) pictured near Zaporizhia. This highly accurate rocket artillery system has been a game changer for Ukraine. (Mil.gov.ua, CC BY 4.0 https://creativecommons.org/licenses/by/4.0, via Wikimedia Commons)

This rear view displays the single MLRS pod of six launchers raised to firing position; the rockets are fired electrically from the armoured cab. (Andrew Kalwitz, U.S. Marine Corps, public domain, via Wikimedia Commons)

A HIMARS launches one of its 227mm rockets. A high level of accuracy is achieved through the use of GPS and inertial guidance. (Dean Johnson, public domain, via Wikimedia Commons)

Stryker ICV

The Striker is a hybrid APC/IFV based on the Canadian LAV III and built for the US Army. Stryker is actually a family of vehicles that equip the rapid deployment Brigade Combat Teams. An eight-wheeled design like the Mowag Piranha, the Stryker comes in ten variants, including: infantry carrier, commander post, ambulance, fire support, engineer, anti-tank guided missile carrier, mortar carrier, reconnaissance vehicle, mobile gun system, and NBC reconnaissance vehicle. The Stryker was first employed operationally in Iraq in 2003, which highlighted a number of issues with the design. The levels of protection had been found inadequate and were improved with the introduction of slat armour and reactive armour tiles to counter the rocket-propelled grenades (RPG) threat. The hull was also re-shaped to a V-configuration to deflect the blast from mines and IEDs. The V-hull shape was not available for all models, however, and the extra weight required further suspension upgrades. The original C7 Caterpillar 350hp engine was superseded by a more powerful 450hp version in later models to improve mobility. Despite the extra protection offered by the V-shaped hull, the tight driver compartment of these versions was to cause problems in operational use prompting a further redesign. There remained concerns about vulnerability to IEDs, although the Stryker proved superior to many other US designs in use in Afghanistan.

The main armament of the Stryker is the M151 remote weapon station mounting a 7.62mm M240 MG, 12.7mm M2 Browning HMG or 40mm Mk19 grenade launcher. The MGS version of the Stryker is armed with a 30mm cannon housed in a Kongsberg Medium Calibre Remote Weapons Station, and trials have been carried out to mount Javelin ATGWs. The vehicle is air-portable by C130 and C17 transport aircraft in keeping with the Stryker Brigades intervention and rapid deployment role. It relies on its speed and manoeuvrability on the conventional battlefield rather than heavy armour. The US has suggested it will supply up to 100 of these capable vehicles to Ukraine, alongside the Bradleys already promised.

A dramatic image of a Stryker crashing through a mud-brick wall in Afghanistan. (Craig Allen)

Specifications	
Model	Stryker ICF
Manufacturer	General Dynamics Land Systems
Country	US
Year	2002
Engine	Caterpillar C7 350hp
Fuel	Diesel
Range	310 miles
Transmission	Allison 3200SP auto
Suspension	8x8
Top Speed	60mph
Armament	7.62mm, 12.7mm Browning M2, 40mm Mk19 GMG, (30mm cannon, MGS)
Capacity	2 crew plus 9 dismounts
Weight	16.4 ton

American soldiers debus from their Stryker troop carrier. Note the jerrycans stowed at the rear and remote weapons station for the 12.7 M2 Browning. (Craig Allen)

Strykers patrol alongside dismounted troops. These examples are equipped with remote weapons stations and a full suite of slat armour. (Craig Allen)

This amusing shot illustrates some interesting features, such as the prominent recovery shackles, wire cutter and additional searchlight mounted on the front quarter. (Craig Allen)

M113 APC

This is the familiar US APC of Vietnam fame, originally developed in the 1960s but modified and updated over the years to keep it current. Replaced in frontline service by the Bradley IFV, significant numbers are still used by the US Army in support roles. The M113 has also been widely supplied to America's allies and is currently in use by Israel, Canada, Australia, Taiwan, South Korea and Pakistan to name but a few. One of the first AFVs to use aluminium armour in its construction, the A2 version came into service in 1979. This enhanced model includes better cooling for the 215bhp 6V-53 Detroit diesel engine an improved torsion bar suspension set up and armoured fuel tanks. This increased the all-up weight to 11,740kg and the A2 lacks the amphibious capability of the earlier version. The later A4 version features an extra road wheel and lengthened hull for more internal space. The M577 variant is simply an M113 series APC with a higher roof to the rear of the driver's position. The hull is still of all-welded aluminium armour providing the occupants with protection from small arms fire and shell splinters. The M577 series was designed as a command post variant with greater room for radio equipment and its operators. Although lacking high levels of protection from IEDs and large-calibre weapons, the M113 series is still useful as a 'battlefield taxi' and has good cross-country performance. In Iraq and Afghanistan, these vehicles received additional slat armour and were sometimes fitted with an

The veteran M113 APC has been in frontline service since the 1960s but has been continually updated and is still useful on today's battlefields. (玄史生, CC BY-SA 3.0 https://creativecommons.org/licenses/by-sa/3.0, via Wikimedia Commons)

armoured turret. In April 2022, the US government agreed to supply 200 M113A2s to Ukraine as part of a military aid package. Australia has also supplied a number of its updated M113AS4s models, while Denmark has sent upgraded M113G4s. Lithuania has also sent M113s, which means the Ukrainian military now has a significant number of these veteran APCs.

Specifications	
Model	M113A2 APC
Manufacturer	United Defence
Country	US
Year	1979
Engine	6V53T, 6-cylinder
Fuel	Diesel
Range	330 miles
Transmission	6-speed manual
Suspension	Torsion bar
Top Speed	42mph
Armament	12.7mm Browning M2
Capacity	2 crew plus 11–15 dismounts
Weight	12.3 ton

M113s heading to Ukraine from Lithuania. These APCs were widely exported to US allies and are still in service all over the world. (Ministry of National Defense of Lithuania, public domain, via Wikimedia Commons)

Troops access the vehicle via the rear loading ramp. Note the simple bench seating and large roof hatch. (Yellowute at English Wikipedia, public domain, via Wikimedia Commons)

Ukrainian troops ride into action on the top of an M113A2 APC mounted with a 12.7 heavy machine gun (HMG) for local protection. (Ministry of Defense of Ukraine)

The Humvee

The High Mobility Multi-Purpose Wheeled Vehicle, more commonly known as the Humvee, was originally developed in the 1980s as a light utility vehicle to replace the jeep. Deployed to Iraq and Afghanistan, it proved fatally vulnerable to RPG strikes and IEDs. Subsequently up-armoured, the M1115 version featured a GPK turret housing an M2 Browning 12.7 HMG or 40mm Mk19 grenade launcher. American-supplied armoured Humvees are currently being used in combat operations by the Ukrainian Armed Forces. Replaced in US service by the MRAP and more recently the JLTV, the Humvee can still play a useful role on the battlefield. They are employed in command-and-control and escort duties, while video footage has shown Humvees providing fire support for dismounted infantry. It appears that those in Ukrainian hands have been armed with 12.7mm Dshk HMG replacing the veteran M2 Browning. This was part of a local modification by the Lviv Machinery and Repair Plant, which originally prepared the vehicles for use by the Ukrainian peacekeeping contingent in Kosovo. Ex-Ukrainian Humvees have also appeared in images of Russian convoys, and it is known that a number of these vehicles fell into the hands of Russian-backed separatists in the Donbas region during fighting in 2014–15. The US supplied a further batch of Humvees in 2020 and has likely supplied more of these vehicles as part of recent military assistance packages.

Specifications	
Model	Humvee
Manufacturer	AM General
Country	US
Year	1983 to present
Armament	12.7 HMG or 40mm GMG
Protection	Welded steel armour
Engine	GEP V8, 6.5 litre 190hp
Fuel	Diesel
Range	250 miles
Transmission	4-speed auto
Suspension	Independent 4-wheel drive, high-low transfer case
Top Speed	70mph
Capacity	1 crew plus 4 dismounts
Weight	3.3 tons

M1115 Humvees fitted with gunners' protection kit (GPK) turrets. Those sent to Ukraine appear to be a mix of gunships and utility models. (Lance Cpl. Brian Marion, public domain, via Wikimedia Commons)

A Humvee in use in Ukraine. This is a standard armoured version, but without the GPK turret. (Mil.gov.ua, CC BY 4.0 https://creativecommons.org/licenses/by/4.0, via Wikimedia Commons)

Left: Ukrainian troops take cover and return fire from behind their Humvee. This appears to be an earlier unarmoured version with a simple shield for the gunner. (moldova.org)

Below: US Humvees arriving by sea at the port of Odesa. This image was probably taken prior to the current conflict, as most supplies now come in by air and land routes. (U.S. Embassy, Kyiv Ukraine, public domain, via Wikimedia Commons)

Chapter 5
German Equipment

Leopard 1

The increasing level of tension between East and West that led to the Cold War made it imperative to re-arm the newly formed German Federal Republic. The Bundeswehr was initially equipped with American M48 Paton tanks but soon began development of its own home-grown design, which became the Kampfpanzer Leopard 1. Relatively lightly armoured but with a focus on mobility and firepower, the new tank was equipped with a licence-built version of the excellent British-designed 105mm L7 rifled gun. The Leopard 1 was developed by Porche, a company responsible for many of Germany's World War Two tanks, but was actually manufactured by KraussMaffei. Powered by an MTU ten-cylinder 819hp engine and relatively light at just over 42 tons, the tank had excellent mobility.

A well-balanced design, it was soon adopted by a number of European armies including those of Italy, Belgium, Norway and the Netherlands. The Leopard 1 was steadily upgraded over the years and from the L1A1 model onwards featured gun stabilisation and a thermal barrel sleeve. Later versions were updated with a ballistic computer, thermal imaging and a laser collimator, and the turret armour was also progressively upgraded. The tank's hull and running gear was used for the Flakpanzer Gepard, which mounts twin 35mm cannons in the anti-aircraft role. The Leopard 1 is still in frontline use in the armies of Brazil, Chile and Turkey, while many more are held in reserve by former users, especially in updated form. Both Germany and Denmark have pledged Leopard 1s to Ukraine in sufficient numbers to equip a whole tank brigade. The German models are Leopard 1A5s, the final version of the tank, which features a larger turret and modernised fire control and thermal imaging equipment. The Leopard 1 may be an older design but is fast and manoeuvrable and has a good gun. Its only weakness is the thin armour, which would make it vulnerable in tank-vs-tank confrontations.

Specifications	
Model	Leopard 1
Manufacturer	KraussMaffei
Country	Germany
Year	1965
Armament	105mm L7 rifled gun, 1 x 7.62mm coax
Protection	Rolled homogeneous armour
Engine	MTU 10-cylinder 819hp
Fuel	Diesel
Range	600km
Transmission	ZF; 4 forward 2 reverse gears
Suspension	Torsion bar
Top Speed	65kph
Capacity	4 crew
Weight	40.2 tons

The Leopard 1 may be an older design but is still effective and is equipped with an accurate and hard-hitting 105mm rifled gun. Note the rounded turret and distinctive rear hull louvres. (Frank Vincentz, CC BY-SA 3.0 <https://creativecommons.org/licenses/by-sa/3.0>, via Wikimedia Commons)

A Leopard 1A5 at speed. The tank is relatively lightly armoured but highly manoeuvrable – note the long-barrelled 105mm gun and the side skirts fitted to later marks. (Frank Vincentz, CC BY-SA 3.0 https://creativecommons.org/licenses/by-sa/3.0, via Wikimedia Commons)

There are many Leopard 1s currently held in storage by former users, and a recent report identified a Belgium warehouse holding over 100 of these tanks. (Seano1, CC BY-SA 3.0 https://creativecommons.org/licenses/by-sa/3.0, via Wikimedia Commons)

Leopard 2

By the early 1970s, both the US and Germany were looking at new designs to compete with the latest generation of Soviet tanks. This was eventually to lead to the development of the American XM1, which became the Abrams. In Germany, the MBT-70 was dropped in favour of redesigning and updating the Leopard 1. The upgrades covered all three vital areas of armour protection, mobility and firepower. A new turret featured almost vertical sides with spaced armour and a prominent rear bustle. The new design of hull was also strengthened to reduced weak spots. This all meant additional weight, but overall mobility was maintained by the use of a 1,500hp engine and improved suspension. Perhaps the biggest improvement was the adoption of the Rheinmetall 120mm smoothbore gun, which was mated to a new fire control system. The Leopard 2 first went into service with the Bundeswehr in 1979, and the initial batches were all fitted with a turret featuring vertical front panels. Later models adopted more angled sloping frontal armour on the turret. The more effective long-barrelled L/55 version of the 120mm gun was also fitted to the later marks of the tank. As well as the new turret and main gun, the Leopard 2A5 version was equipped with side skirts and an appliqué armour package. The latest version of the tank is the Leopard 2A7, which features further enhancements including improved optics and new ammunition capable of penetrating the latest Russian tanks.

The news that Germany, Poland and other European allies are to send some of their Leopard 2s to Ukraine begs the question of why this tank is in such high demand. The answer perhaps lies not just in the capabilities of the Leopard 2 but in its availability. Produced in significant numbers and used by several European nations, there are lots of these tanks that could potentially be donated. With a frugal diesel engine and utilising standard NATO ammunition, it is also relatively easy to supply and maintain. The Leopard 2 was originally designed to counter the Russian T-90, the tank it will face in Ukraine, so in many ways is ideal for this conflict.

The Leopard 2 is obviously very capable, but why is it a better fit for the Ukrainian military than the Challenger, Abrams or even the French Leclerc? The Challenger is a very capable tank and better protected than the Leopard 2, however, this British tank has only been produced in relatively small numbers and uses non-NATO standard ammunition for its 120mm rifled gun. Then there is the Abrams recently offered by the Americans, which is one of the most advanced and combat-tested tanks in the world. The Abrams may have proved its worth in battle but is complex, difficult to maintain in the field and uses an extremely thirsty gas-turbine engine. It will also take time to train up the crews and get the promised 31 Abrams M1A2s all the way from the United States. Then there is the wild card, the Leclerc, which is available now the French have stated they are open to the idea of providing heavy armour. The Leclerc is a good design and uses NATO standard ammunition, however, like the

Challenger, it has only been produced in small numbers. The Leopard 2, on the other hand, is used by 13 European countries with potentially hundreds available. They are also much closer to hand with good access to spare parts. Ideally, the Ukrainians need to concentrate on one design of MBT to ease both training and maintenance and the Leopard 2 clearly makes the most sense.

Germany, Poland, Finland and Canada are all providing some Leopard 2s, and at the time of writing the first batch of Polish Leopards has arrived in the country. This means the Ukrainians finally have the prospect of putting together an effective modern tank force. Many of these tanks will be older A4 models coming from storage but will still be effective against the Russian armour. At least some of the German tanks are reported to be later A6 models, more than a match for the latest Russian T90s.

Specifications	
Model	Leopard 2
Manufacturer	KraussMaffei
Country	Germany
Year	1979–to present
Armament	Rheinmetall 120mm L/55 smoothbore gun
Protection	Rolled homogeneous armour
Engine	MTU MB 873 V12 1.479hp
Fuel	Diesel
Range	340km
Transmission	Renk HSWL 354 hydro-mechanical
Suspension	Torsion bar
Top Speed	70kph
Capacity	4 crew
Weight	62.3 tons

A Leopard 2A5 in service with the Bundeswehr. Note the steeply sloped frontal armour on the turret and banks of smoke dischargers. (Boevaya mashina, CC BY-SA 4.0 https://creativecommons.org/licenses/by-sa/4.0, via Wikimedia Commons)

A Leopard 2A7 is the latest version of the tank pictured. Most of the donated tanks are likely to be older Leopard 2A4s. (Fric.matej, CC BY-SA 4.0 https://creativecommons.org/licenses/by-sa/4.0, via Wikimedia Commons)

The Leopard 2A4. Note the vertical front turret armour. These models are fitted with the older L/44 version of the 12mm smoothbore gun. (Jonathan G. Seow H. C. CC BY-SA 4.0 https://creativecommons.org/licenses/by-sa/4.0, via Wikimedia Commons)

The ceremony held to mark the arrival of the first Leopard 2 tanks donated by Poland. (Kmu.gov.ua, CC BY 4.0 https://creativecommons.org/licenses/by/4.0, via Wikimedia Commons)

Gepard Anti-Aircraft System

The Gepard is clearly influenced by the Flakpanzers of World War Two fame and combines twin turret-mounted cannons with the chassis of the Leopard 1 tank. The large turret is electrically powered and has 360-degree traverse while elevating the twin cannon to almost 90 degrees. It is equipped with two radar dishes, one for search and the other for tracking. Later models also feature a laser rangefinder, and the electronics have been steadily updated. The armament consists of a pair of 35mm Oerlikon auto-cannon's firing frangible armour piercing discarding sabot (FAPDS) rounds for use against aircraft. There is also armour-piercing ammunition available for use in the ground role. With each of the twin cannons firing at a rate of 550rpm, the Gepard can produce an impressive amount of firepower. The search and tracking radars are effective out to a range of 15km and are mated to a digital fire control system. The system is deadly against combat aircraft and helicopters flying at low to medium altitudes. In Ukraine, the Gepard has proved equally effective when employed against Russian cruise missiles and Iranian-produced Shahed kamikaze drones. Well protected and with excellent mobility due to its tracked chassis, it is highly valued in the air-defence role. So far, the German government has provided Ukraine with 37 Gepards and is working to supply more by refurbishing further examples. These are coming from reserve stocks, as the Bundeswehr has replaced the system with the Wiesel 2 Ozelot, armed with Stinger missiles. Both Belgium and the Netherlands also hold stocks of Gepards, which could be donated in future. There has been some concern for the supply of ammunition, as the specialised anti-aircraft 30mm rounds are manufactured mainly by the neutral Swiss. German automotive and arms manufacturer Rheinmetall has therefore stepped in to begin production and ensure continuity of supply.

The Flakpanzer Gepard uses the chassis and running gear of the Leopard 1 tank. Note the twin radar dishes and armoured sponsons for the twin 35mm cannon. (Hans-Hermann Bühling, CC BY-SA 3.0 http://creativecommons.org/licenses/by-sa/3.0/, via Wikimedia Commons)

German Equipment

Specifications	
Model	Flakpanzer Gepard
Manufacturer	KraussMaffei
Country	Germany
Year	1973
Engine	MTU MB Ca M500 830hp
Fuel	Diesel
Range	600km, road
Transmission	ZF 4HP250; 4 forward 2 reverse gears
Suspension	Torsion bar
Top Speed	65kph
Armament	Twin 35mm Oerlikon canon
Protection	Rolled homogeneous armour
Capacity	3 crew
Weight	47.3 tons

This side view gives an indication of the size of the Gepard turret. Note the grenade launchers mounted on the turret sides. (Hans-Hermann Bühling, CC BY-SA 2.0 DE https://creativecommons.org/licenses/by-sa/2.0/de/deed.en, via Wikimedia Commons)

A Gepard pictured during firing practice with the twin Oerlikon 35mm autocannons. Note the ejected casings from the spent ammunition. (Bundeswehr-Fotos, CC BY 2.0 https://creativecommons.org/licenses/by/2.0, via Wikimedia Commons)

PzH 2000 SPG

The Panzerhaubitze 2000 is a German-built 155mm self-propelled Howitzer built by KraussMaffei Wegmann and Rheinmetall for the Bundeswehr. Initially designed in the 1980s, it uses the chassis and running gear of the Leopard 2 tank. The Rheinmetall 155mm L52 gun is mounted in a 360-degree armoured turret and features an autoloader. It can use either modular or bagged charges and has a range of some 36km with standard ammunition. This can be extended using nosebleed and rocket-assisted munitions to ranges in excess of 50km. The firing is controlled by a digital multiple round simultaneous impact system. The speed and accuracy of the PzH 2000 is truly impressive. The mechanical autoloader allows rates of fire of up to 20 rounds in under two minutes. The barrel can then be locked down automatically, allowing rapid moves to a fresh location. The tracked chassis has excellent mobility and can reach speeds of up to 45km across-country. The PzH 2000 has been exported to Denmark, Italy and the Netherlands, amongst other countires, and the Dutch used them operationally in Afghanistan. So far, Germany has supplied seven of these self-propelled guns to Ukraine as well as stocks of ammunition; the Dutch five; and Italy a further 22.

Specifications	
Model	PzH 2000 SPG
Manufacturer	KMW & Rheinmetall
Country	Germany
Year	1998
Armament	Rheinmetall 155mm L52 gun
Protection	Welded steel armour
Engine	MTU MT881 1000hp
Fuel	Diesel
Range	420km
Transmission	Renk 284C
Suspension	Torsion bar
Top Speed	67kph
Capacity	3 crew
Weight	55.8 tons

A PzH 2000 SPG opens fire with its 155mm Howitzer, producing an impressive muzzle blast. Note the rear hatch, which facilitates loading and ejecting shells. (CroPatriot, public domain, via Wikimedia Commons)

This side view shows the large size of the turret relative to the hull, which is derived from the Leopard 2. Note the length of the 155mm gun barrel. (mezha.media)

The Marder IFV

The Marder came into service with the Bundeswehr in the 1970s and was one of the first Western-produced IFVs. It is a conventional design using a welded steel hull with a two-man centrally located turret. The turret mounts a Rheinmetall 20mm automatic cannon with coaxial 7.62mm MG plus a single MILAN missile launcher. The Marder has a three-man crew and can accommodate five dismounts in the main troop compartment. This is accessed via a rear loading ramp and there are an additional three top hatches. The Marder was designed to keep up with the Leopard 1 tank so has good cross-country mobility. Early versions could reach speeds of up to 75kph, but this has been reduced with the fitting of additional armour. Gun ports were originally located in the hull sides, but these were obscured by the later appliqué armour packages. Marders were deployed to Kosovo with peacekeeping force KFOR and also saw some limited combat in Afghanistan. A modern replacement, the Marder 2, never got beyond the prototype stage, and instead the Marder 1 was steadily upgraded. The A2 version featured a new sighting system including a thermal imager and improved suspension. The A3 Marder received enhanced frontal armour and there were A4 and A5 variants that mainly saw improvements to the communications fit and added blast protection. These later versions were only produced in fairly limited numbers. The Marder continues to soldier on with the Bundeswehr, although the new Puma IFV is gradually replacing it. In January 2023, the German government pledged to supply around 40 Marders to Ukraine. The Marder may not be the most modern of IFVs but is still effective and well able to keep up. It also offers some useful firepower and will be a valuable addition to the Ukrainian armoury.

Specifications	
Model	Marder 1A2
Manufacturer	Rheinmetall
Country	Germany
Year	1971
Engine	MTU MB 833 591hp
Fuel	Diesel
Range	550km
Transmission	Renk 4-speed HSWL hydrostatic gearbox
Suspension	Torsion bar
Top Speed	75kph (65kph for later variants)
Armament	20mm cannon with 7.62mm Coax MG plus Milan Launcher
Protection	Welded steel with appliqué armour package
Capacity	3 crew plus 5 dismounts
Weight	28.5 ton

The Marder may be an older design but is still in service and still considered effective on the modern battlefield. Note the turret-mounted grenade launchers and hosing for the 7.62mm coax MG. (funky1opti from Wolfsburg, Deutschland, CC BY 2.0 https://creativecommons.org/licenses/by/2.0, via Wikimedia Commons)

Troops dismount from the rear loading ramp. The Marder can then give them fire support with its 20mm autocannon. (Bundeswehr-Fotos, CC BY 2.0 https://creativecommons.org/licenses/by/2.0, via Wikimedia Commons)

The Marder is highly manoeuvrable and can reach speeds of up to 65kph on roads. (mezha.media)

Chapter 6
French Equipment

AMX 10 RC

The AMX 10 RC is an armoured reconnaissance vehicle and with its powerful 105mm main gun, can be considered as a wheeled light tank. It has been in service with the French Army since the 1980s and remains in use by its armoured cavalry regiments. The hull and turret are constructed from welded aluminium, so the vehicle is relatively lightly armoured. With its 6x6 wheeled chassis, the AMX 10 has good all-round mobility and is fully amphibious. The hydropneumatic suspension means that ground clearance can be altered to suit the terrain when operating off-road. It also features a central inflation system and run-flat tyres. The three-man turret mounts a 105/47 F2 MECA 105mm rifled gun that can fire APFSDS, HE, and HEAT rounds out to a range of over 2,000m. The gun lacks stabilisation so is unable to fire accurately when on the move, but it does have a high first round hit capability. Although not considered effective against the latest generation of MBTs, it still represents a threat to older Russian tanks and armoured vehicles. Power is derived from a Baudouin GF-11SX 280hp diesel engine driving through a preselector gearbox. This provides four forward and four reverse gears and is fitted with a torque converter.

The AMX 10 RC had had several upgrades over the years, including hardened steel appliqué armour, thermal sights and improvements to the 105mm gun including a new muzzle brake. The modernised AMX 10 RCR version features the SIT battlefield management system additional armour protection and an auto-gearbox. Enhanced battlefield communications and improved NBC protection is also fitted. The AMX 10 RC saw operational service in the first Gulf War and has since been employed in Kosovo, Afghanistan and Mali. It is currently being replaced by the new Jaguar reconnaissance vehicle in French service. So far, around 40 of the earlier RC versions have been provided to Ukraine. These should prove useful for general fire support and taking on the Soviet-era Russian armour.

Specifications	
Model	AMX 10 RC
Manufacturer	Nexter Systems
Country	France
Year	1981
Armament	105/47 F2 MECA 105mm rifled gun
Protection	Welded aluminium, plus hardened steel armour
Engine	GF-11SX 280hp
Fuel	Diesel
Range	1,000km
Transmission	4-speed electro-magnetic
Suspension	6x6 hydropneumatic
Top Speed	85kph
Capacity	4 crew
Weight	15.8 tons

The AMX10 is effectively a wheeled tank and mounts a powerful 105mm gun. Note the prominent turret bustle and rear-mounted engine louvres. (Officer communication du 4e RCh, CC BY-SA 4.0 https://creativecommons.org/licenses/by-sa/4.0, via Wikimedia Commons)

With its wheeled chassis and hydropneumatic suspension, the AMX10 has good mobility and can reach speeds of up to 85kph. (Lance Cpl. Juanenrique Owings, public domain, via Wikimedia Commons)

The AMX 10 has seen action in Kosovo, Afghanistan and Mali and is seen here on an exercise in a desert setting. (United States Marine Corps, public domain, via Wikimedia Commons)

Ceaser SPG

The Ceaser is a self-propelled Howitzer mounted on either a 6x6 or 8x8 truck chassis and is still in use by the French Army. First developed in the 1990s, it went into service in 2008 and has been used operationally in Iraq, Afghanistan and Mali. The 155mm gun can fire all standard NATO ammunition types and has an effective range of some 40km. A Thales fire control system, along with GPS navigation, ensures accuracy. Before firing, a rear-mounted hydraulic platform or spade is deployed to counter recoil vibrations. In the original form, the weapon itself uses a semi-automatic loading system and is mounted on a Renault Sherpa 6x6 chassis. Nexter have also developed Unimog 6x6 and Tatra 8x8 versions for export, and the Ceaser is also operated by Denmark, Indonesia and Saudi Arabia. The standard French Army version is powered by a Renault 6-cylinder 245hp diesel engine driving through a six-speed manual gearbox. The crew cab offers protection from small arms and shell fragments and can be up-armoured if required. The Ceaser can be brought into action rapidly and fires six rounds within two minutes, giving it a useful 'shoot and scoot' ability. The French have so far delivered 18 Ceasers plus ammunition and Denmark has provided a further 19 from its own stock.

Specifications	
Model	Ceaser truck-mounted Howitzer
Manufacturer	Nexter Systems
Country	France
Year	2008
Armament	155mm Howitzer
Protection	Armoured crew cab
Engine	Renault 6-cylinder 245hp
Fuel	Diesel
Range	600km
Transmission	6-speed manual (auto available)
Suspension	6x6 hydropneumatic
Top Speed	50kph
Capacity	5/6 crew
Weight	17.7 tons

The Ceaser truck-mounted artillery system is already in use in Ukraine, with the weapons coming from France and Denmark. (Aljazeera)

This shot illustrates the use of the rear-mounted spade that stabilises the weapon for firing. (defensehere.com)

The Ceaser has good cross-country mobility and can come in and out of action rapidly, which gives it a useful 'shoot and scoot' capability. (Mil.gov.ua, CC BY 4.0 https://creativecommons.org/licenses/by/4.0, via Wikimedia Commons)

VAB APC

The Véhicule de l'Avant Blindé (Armoured Vanguard Vehicle or VAB) is a light wheeled APC manufacture by the Renault company. Designed to complement the AMX10 armoured reconnaissance vehicle, the VAB is lightly armoured but highly manoeuvrable. In fact, the protection levels are only sufficient to resist small arms fire and shell fragments. However, later versions did receive a composite armour package that increased protection to withstand strikes from 12.7mm rounds. The standard version of the VAB featured a 7.62mm MG mounted in an open turret, while a more heavily armed version was fitted with a 12.7 HMG. Later variants of the VAB were equipped with a remote weapon station mounting a 12.7mm HMG or 20mm cannon. The vehicle is powered by a Renault MIDS six-cylinder 320hp turbo-diesel engine and was originally fitted with a manual transmission. These were later upgraded to automatic, and the VAB is equipped with a four-wheel drive system and was originally fully amphibious. A two-man crew and up to ten dismounts can be accommodated on benches in the spacious rear body. Twin armoured doors allow access to the troop compartment, while the crew have small side-mounted doors. There are also a pair of roof-mounted hatches in addition to the forward gunner's hatch. First adopted by the French Army in the mid-1970s, the VAB has been steadily updated over the years. The improvements include upgraded engines, MEXAS composite armour, central tyre inflation and remote weapon stations. It's worth noting that up-armoured versions lose their amphibious capability due to the increased weight. The VAB was produced in several different variants, including ambulance, electronic warfare, engineer, NBC recce and anti-tank versions. In August 2022, the French government announced it was transferring significant amounts of the these light APCs to Ukraine.

The Véhicule de l'Avant Blindé (Armoured Vanguard Vehicle or VAB) is a light wheeled APC in service with the French Army. Note the boat-shaped hull, small crew doors and armoured flaps for the windscreens. (AlfvanBeem, CC0, via Wikimedia Commons)

This rear view displays the twin armoured troop doors and roof hatches. Note the high ground clearance of the vehicle. (David Monniaux, CC by-SA 3.0 http://creativecommons.org/licenses/by-sa/3.0/, via Wikimedia Commons)

Specifications

Model	VAB APC
Manufacturer	Renault Trucks
Country	France
Year	1976
Armament	7.62mm MG, 12.7mm HMG or 20mm cannon
Protection	Welded steel armour
Engine	Renault MDIS 320hp
Fuel	Diesel
Range	1,200km
Transmission	Auto
Suspension	Fully independent 4x4
Top Speed	110kph
Capacity	2 crew plus 10 dismounts
Weight	13.8 tons

The capacious rear body can house up to ten fully equipped infantrymen on the inward-facing bench seats. Note the twin roof hatches. (Jastrow, CC by 2.5 https://creativecommons.org/licenses/by/2.5, via Wikimedia Commons)

A VAB in service with the Ukrainian Air Assault Forces. Note the ammunition chute for the 12.7mm HMG and the front mounted recovery points. (mil.in.ua)

Chapter 7
Polish Equipment

PT-91 Twardy

The PT-91 is essentially a modernised and enhanced version of the Russian/Soviet T-72M1, which is the export model of the tank. Development work began in the early 1990s, facilitated by the fact the Poles had already built T-72s under licence. The upgrades centred on the areas of protection, fire control and mobility and were all home-produced. To improve survivability against ATGWs, the PT-91 is equipped with Polish-designed ERAWA tiles. These are a much neater fit than the reactive armour tiles typically seen on Russian T-72B3s. The Twardy has the same 125mm main gun as the Russian tank but features a new design of auto-loader capable of up to 10rpm. The SKO-1M Drawa fire control system includes a ballistic computer plus laser range finder, and the gunner is provided with a day/night sight with thermal imaging. To counteract the extra weight of the armour, a more powerful engine is fitted; this is the Polish-built S-12U 12-cylinder engine that achieves 850hp, a modern development of the old Soviet V-46 powerplant. The latest variants feature the even more powerful S-1000 1,000hp engine. Poland is sending up to 60 of these modernised T-72 variants to Ukraine.

The PT-51 Twardy is a Polish-built upgrade of the Soviet T-72M1, which includes a number of home-produced features including its ERA tiles and modern fire control system. (Maciej Baranowski, public domain, via Wikimedia Commons)

Specifications

Model	PT-51 Twardy
Manufacturer	Bumar-Labedy
Country	Poland
Year	1995
Armament	125mm 2A46 gun with autoloader
Protection	Laminate/composite, ERAWA tiles
Engine	S-12U 12-cylinder 850hp
Fuel	Diesel
Range	650km
Transmission	Manual
Suspension	Torsion bar
Top Speed	70kph
Capacity	3 crew
Weight	45.9 tons

Right: The PT-51 uses Polish-designed ERAWA tiles, which are a much neater fit than the equivalent Russian tiles. (Michal Derela, CC by-SA 4.0 https://creativecommons.org/licenses/by-sa/4.0, via Wikimedia Commons)

Below: A pair of PT-51s on an exercise in Poland. Note the low silhouette and banks of turret-mounted smoke dischargers. (gagadget.com)

The PT-51 represents a significant improvement over the original T-72M1 and should prove useful to the Ukrainians, who are already familiar with these ex-Soviet tanks. (Bumar Labedy via Bulgarianmilitary.com)

Krab SPG

The Krab 155mm SPG is manufactured in Poland by Huta Stalowa Wola (HSW) for use by the country's armed forces. The vehicle combines features from several different designs: the chassis comes from the South Korean K9 SPG, and this is mated to a British AS90 turret with a French main gun. The .52-calibre barrels were originally made by Rheinmetall, but more recently they have been manufactured within Poland. The Topaz electronic fire control system used for gun laying is also home produced. The 155mm gun uses standard NATO ammunition and has a range of 30km, which can be extended using precision ammunition. The Krab is powered by a German-built eight-cylinder diesel engine developing 1,000hp and uses an Allison four-speed transmission. The tracked chassis features hydropneumatic suspension and gives the vehicle excellent cross-country mobility. Poland has supplied 18 Krabs to Ukraine, and they have already seen action in the counterattacks around Kharkiv. The Ukrainians have ordered up to 60 more of these effective self-propelled howitzers, but it will take time for them to be manufactured and delivered.

Specifications	
Model	Krab SPG
Manufacturer	Huta Stalowa Wola (HSW)
Country	Poland
Year	2006
Armament	Nexter 155mm .52-calibre gun
Protection	Welded steel armour
Engine	MTU MT881Ka-500 1000hp
Fuel	Diesel
Range	400km
Transmission	Allison; 4 forward 2 reverse
Suspension	Hydropneumatic
Top Speed	60kph
Capacity	5 crew
Weight	48 tons

The Krab SPG combines features from several other designs but is a capable modern Howitzer that can fire all types of 155mm standard NATO ammunition. (MilitaryLeak.com)

The Krab is already in use in Ukraine, with many more on order from the Polish manufacturer. Note the long 155mm barrel with box-shaped muzzel brake. (MilitaryLeak.com)

Chapter 8
Other Nations

The M-55S Tank

It was recently reported that Slovenia has given Ukraine a total of 28 of its M-55S tanks. These are modernised versions of the veteran T-55, and were held in storage after being replaced by the more modern M-84. The deal with Ukraine also apparently involved Germany supplying Slovenia with a consignment of 40 military trucks. The number of M-55s provided is sufficient for a single tank battalion, and it appears they have been issued to the Ukrainian 47th Assault Brigade.

The M-55S in essence is a deeply modernised version of the old Soviet T-54/55 to make it viable on the modern battlefield. The conversion work was carried out in the late 1990s by Israeli company Elbit and RAVNE of Slovenia to modify 30 of the ageing T-55s. The biggest change was to exchange the old 100mm gun for the British 105mm Royal Ordnance L7 as the main armament. The turret and hull were then covered in modern reactive armour panels, significantly altering the profile of the tank. They also received a digital ballistic computer, gun stabilisation and a laser rangefinder with the commander's position benefiting from a new COMTOS-55 sight. A more powerful V12 diesel engine and rubber/metal tracks improved mobility, and added protection was provided by a laser radiation detector linked to the tank's smoke grenade launchers. The result of all this effort was a hybrid of Soviet-designed turret and hull with a modern Western-supplied gun. Undoubtedly, it is the fitting of that new armament that makes these tanks effective. The 105mm Royal Ordnance L7 gun is a successful design formally fitted to the British Centurion and German Leopard 1. It is also compatible with a wide range of modern ammunition types. These include armour-piercing rounds capable of penetrating a Russian T-72 at considerable ranges. The gun is rifled, features a distinctive thermal jacket and has a maximum range of some 4,000m.

At first glance, these older-generation tanks might appear of limited value to the Ukrainians in their current fight. Nevertheless, the combination of the uprated and modernised T-55 hull with an effective gun could prove to be a winning combination. Properly handled, they should be capable of taking on Russian tanks such as the T72 and give vital support to the hard-pressed infantry.

Specifications	
Model	M-55S
Manufacturer	Elbit and RAVNE
Country	Slovenia
Year	1998
Armament	105mm L7 rifled gun, 1 x 7.62mm Coax MG, 1 x 12.7 MG
Protection	100mm rolled steel armour plus ERA
Engine	V12 800hp
Fuel	Diesel
Range	350km
Transmission	5-speed synchromesh
Suspension	Torsion bar
Top Speed	50kph
Capacity	4 crew
Weight	38 ton

The Slovenian M-55S tank combines a Soviet-era design with a modern British-designed 105mm L7 gun. (https://creativecommons.org/licenses/by/3.0 via Wikimedia Commons)

The M-55S is based on the classic Cold War-era T-54/55 tank produced in huge numbers in the 1950s. (Eric Kilby, CC by-SA 2.0, https://creativecommons.org/licenses/by-sa/2.0, via Wikimedia Commons)

The add-on reactive armour plates significantly alter the profile of this Cold War tank. Note the thermal sleeve on the 105mm main gun. (MilitaryLeak.com)

Bushmaster PMV

The Bushmaster is a wheeled light APC featuring an armoured body that is fully protected from mines and IEDs. Designed by Australian Defence Industries in the mid-1990s, it was accepted by the Australian military as its PMV. The V-shaped monocoque hull is designed to deflect blast, while the armour is proof against small arms and shell fragments. There is an additional appliqué armour package that can boost the protection levels, if required. A forward-located gun ring was originally fitted for a 7.62mm MG, although later updates included a remote weapon station for a heavier 12.7mm HMG. The vehicle is powered by a Caterpillar six-cylinder turbo-diesel engine developing 300hp and driving through a ZF six-speed transmission. Combined with its fully independent coil-sprung suspension, this gives the vehicle excellent mobility with a maximum road speed of 100kph. The four-wheel drive system and high ground clearance means it is also very capable off-road. Along with a two-man crew, the Bushmaster can carry up to eight infantry dismounts and troop access is via the large rear door. Fully air-portable, the vehicles can be lifted by C130 and C17 transport aircraft. The Bushmaster can be configured for several different roles, including basic troop transport, command, ambulance, assault pioneer and mortar variants. The vehicle was first used operationally in Iraq and saw service in Afghanistan with both Australian and Dutch forces. A small number of Bushmasters were also acquired by the British as strike vehicles for use by Special Forces in Iraq.

An initial batch of 20 Bushmasters was donated to Ukraine by the Australian government in April 2022. These comprised troop carriers with remote weapon stations and a smaller number of the ambulance version. Additional supplies of these vehicles made since means that the Ukrainian military currently has around 90 of these light armoured troop carriers in frontline use.

The Bushmaster PMV is a fully mine- and blast-protected light troop carrier currently in service with Australian forces. Note, this is the earlier version with a simple gun ring for a 7.62mm MG. (Hpeterswald, CC by-SA 3.0, https://creativecommons.org/licenses/by-sa/3.0, via Wikimedia Commons)

Specifications	
Model	Bushmaster PMV
Manufacturer	ADI (Thales)
Country	Australia
Year	1997
Armament	12.7mm HMG in remote weapon station
Protection	V-shaped monocoque hull
Engine	Caterpillar 3126E 7.2 litre 300hp
Fuel	Diesel
Range	350km
Transmission	ZF-6-speed
Suspension	Fully independent progressive coils
Top Speed	100kph
Capacity	2 crew plus 8 dismounts
Weight	15 ton

A Bushmaster in use in Ukraine. Note, this vehicle is fitted with a remote weapon station mounting that appears to be 7.62 PK MG. (Світлана Кирган/Інформаційне агентство АрміяInform, CC by-SA 4.0 https://creativecommons.org/licenses/by-sa/4.0/, via Wikimedia Commons)

The blast-attenuating rear seating allows for the carriage of up to eight infantry dismounts. Note the AT4 shoulder-launched ATGW stored above the seat backs. (Kitmaker.net)

ACSV

The Armoured Combat Support Vehicle (ACSV) is based on the LAV built by General Dynamics and is replacing the M113 and LAV II Bison in Canadian service. The ACSV features a raised rear hull and is intended to fulfil a number of roles including command, recovery, reconnaissance and armoured ambulance. The 8x8 wheeled chassis comes from the latest LAV 6, which is replacing earlier versions of the LAV. Heavily armoured, the ACSV features a double-skinned V-shaped hull for enhanced blast protection. The extra weight is offset by the powerful caterpillar C7 450hp engine, which drives through a seven-speed ZF auto transmission. Troop access is via a rear-mounted loading ramp and there is space for up to eight dismounts. The Canadians have committed to buying up to 360 of the new ACVS to fill support roles and have diverted 39 from the initial tranche to Ukraine. These began arriving in the country in late 2022 and have recently been spotted in use on the frontline.

Specifications	
Model	ACVS
Manufacture	General Dynamics
Country	Canada
Year	2019
Engine	Caterpillar C7 450hp
Transmission	ZF 7-speed auto
Fuel	Diesel
Range	600km
Suspension	Independent hydropneumatic
Top Speed	62kmh
Armament	7.62mm or 12.7mm MG
Capacity	3 plus 8 dismounts
Weight	45 ton

The Armoured Combat Support Vehicle (ACSV) features a raised rear hull and forward-mounted commander's cupola. It can be used as an APC or for various support roles. (https://www.canada.ca/content/dam/dnd-mdn/images/procurement/projects/2021-07-12-acsv-vbsc.jpg)

The ACVS is based on the LAV 6, which is currently replacing earlier versions of the vehicle in Canadian service. (Oshkoshdefense.com)

An ACVS at speed in the armoured ambulance role. Note the spare tyre stowed on the front glacis plate. (Government of Canada, CC by-SA 4.0 https://creativecommons.org/licenses/by-sa/4.0, via Wikimedia Commons)

Senator APC

These light armoured vehicles are manufactured by Canadian company Roshel and are similar in concept to the British Foxhound. Built on the chassis of the Ford F-550 truck, the armoured body offers protection from small arms fire and shell fragments. The vehicle can also be fitted with an armoured turret for mounting a 7.62mm or 12.7mm machine gun. The Senator features a four-wheel drive system and is fitted run-flat tyres. While originally conceived as a SWAT team vehicle and for peacekeeping duties, the Senator can also serve as a light APC. It may not be a fully mine-protected vehicle like the Foxhound but is still useful as a battle taxi. Canada initially supplied eight of these vehicles to Ukraine, which were employed by the country's Border Guards. In January 2023, it was announced that a further 200 Senators would be provided, with the makers Roshel ramping up production accordingly.

Specifications	
Model	Senator APC
Manufacture	Roshel
Country	Canada
Year	2020
Engine	6.7-litre power stroke 330hp
Transmission	6-speed TorqShift auto
Fuel	Diesel
Range	NA
Suspension	Coil springs, fully independent
Top Speed	NA
Armament	7.62 and 12.7mm MGs can be fitted
Capacity	2 plus 10 dismounts
Weight	8 ton

Left: Canadian-supplied Senator APCs in use by the Ukrainian Border Guards; these examples lack turrets or armaments. Note the front-mounted winch. (suspilne.media.com)

Below left: This manufacturer's image shows the Senator with its armoured turret. This appears to be similar to the OGPK turret fitted to US Humvees. (Roshel Smart Armored Vehicles)

Below right: This view of a Ukrainian Senator displays the armoured rear door for troop access. Note the ladder for access to the roof. (LeMonde)

CV90 IFV

In early 2003, the Swedish defence minister promised Ukraine up to 50 of these IFVs. The CV90 was actually a joint project between Hagglunds and Bofors to produce a series of armoured vehicles for the Swedish Army. The IFV version mounts a Bofors 40mm autocannon in a two-man turret and was officially adopted in 1993. Norway also ordered the vehicle in improved Mk1 form with a 30mm turret. The CV90 features a tracked armoured hull, which offers protection from 14.7mm projectiles and up to 30mm rounds on the frontal arc. The vehicle can also be fitted with additional appliqué panels and slat armour to improve protection levels when required. Powered by a Scania 14-litre diesel engine driving through an auto transmission, the CV90 has good all-round mobility. In addition to the three-man crew, a full eight-man infantry section can be carried with access via a conventional rear ramp. In 2008, a Norwegian contingent took its CV90s to Afghanistan, where they received their first combat use against the Taliban. Denmark also employed CV90s in Helmand Province in 2010, with a number suffering damage from powerful IEDs. Operational experience has led to several upgrades in later marks of the vehicle, with improved protection levels and more powerful engines. In Swedish service, variants include command, air defence, mortar, recovery and electronic warfare versions, and light tank demonstrators have also been produced equipped with a 120mm smoothbore gun. The CV90 is a well-proven and widely exported design that offers a similar capability to the US Bradley and British Warrior IFVs.

The CV 90 is a Swedish designed infantry fighting Vehicle armed with a powerful Bofors 40mm Gun in a two-man turret. (Anders Lageråsderivative Sonaz, CC BY-SA 3.0 <https://creativecommons.org/licenses/by-sa/3.0>, via Wikimedia Commons)

Specifications	
Model	CV90 IFV
Manufacturer	Hagglunds/Bofors
Country	Sweden
Year	1993
Armament	40mm Bofors L/70 auto-cannon, 35mm/50mm Bushmaster cannon, 7.62mm coaxial MG
Protection	welded steel armour
Engine	Scania DS14 550hp
Fuel	Diesel
Range	320km
Transmission	Perkins auto
Suspension	Torsion bar
Top Speed	70kph
Capacity	3 crew plus 8 dismounts
Weight	23 tons (original variant)

Swiss Army CV90s on exercise, note the hinged side skirts and turret mounted grenade launching clusters. (Kecko, https://creativecommons.org/licenses/by/2.0, via Wikimedia Commons)

'Armourgeddon' in Ukraine

In the lead-up to the invasion of Ukraine, in February 2022, Russia was recognised as having one of the largest and most effective tank armies in the world. Consisting of modernised T-72s, T-80s and the latest T-90s operating in combined arms groups, they were expected to decimate all that came before them. However, exactly the opposite happened, and the blasted hulks of Russian tanks became a daily feature of our news reports. To the surprise of most Western observers, the armoured thrust on Kyiv was halted in its tracks then turned back – so what went wrong for Russia?

The Tanks

The bulk of the Russian armour is made up of T-72s and T-80s, both of which date from the 1970s. These tanks have been progressively upgraded over the years, including the fitting of reactive armour panels and improved sights and fire control systems. However, the T-80 performed poorly in the first Chechen War of 1994, with many lost to shoulder-fired anti-tank missiles. These were often launched from high buildings targeting the vulnerable upper areas of the tank. This form of 'top attack' is therefore nothing new and not restricted to the latest generation of anti-armour weapons. What is perhaps new is the precision and enhanced effectiveness of anti-tank missiles such as Next Generation Light Anti-Tank Weapon (NLAW) and Javelin. The turret cages fitted to some Russian tanks may have been intended to counter the top attack mode of these weapons but in practice have proved ineffective. Another lesson of the Chechen War was that tanks are very vulnerable in built-up areas if unsupported by infantry. This seems to have been forgotten in the current conflict, which has allowed Ukrainian tank-hunting teams to get dangerously close, and the destruction wrought on so many Russian tanks with turrets blown clean off is the result. This occurs when ammunition held in the hull violently 'cooks off' in spectacular fashion. Storing the spare ammunition in a hull-mounted carousel and the use of an autoloader allows for a smaller turret, however. This is a trade-off that gives the tank a reduced silhouette but has proven an Achilles heel in the recent fighting. The T-90, a third-generation version of the T-72 from the 1990s, meanwhile, is a far more modern design, featuring composite armour, the latest Kontakt 5 reactive armour and the Shtora active defence system against anti-tank guided missiles (ATGMs). However, it is only available in smaller numbers, and it has proved equally vulnerable to modern anti-tank weapons if left unsupported. The T-72 is arguably the best performer. It is the most widely used battle tank in the world and is also used by the Ukrainians. It was employed successfully in the counteroffensive in the East alongside modernised versions of the T-80, which begs the question of whether it is the Russian tanks or the tactics that are at fault.

Tactics

Released footage of Russian armoured units in training before the war featured tanks and infantry fighting vehicles charging across open country closely supported by Russian aircraft. This is the kind of combined arms manoeuvre warfare that is enshrined in Russian military doctrine. Nevertheless, this is not what was seen in February 2022, when the armoured spearheads invaded Ukraine. The massed columns of tanks and armoured vehicles were largely road-bound and attacked on several widely separated axis. It also appeared that the Russian battalion groups were short of infantry support and the armour was often left to operate alone. This made the columns vulnerable to attack while close air support from Russian aviation was surprisingly lacking. The armour was poorly handled, driving nose-to-tail and failing to disperse during halts. As in Chechnya, the tanks were especially vulnerable in urban areas often falling victim to Ukrainian tank-hunting teams. Road-bound columns were sometimes engaged by a single opposing tank firing from a concealed position. All of this speaks to poor crew training and a lack of competent leadership at the tactical level. Instead of being used together, armour, infantry and artillery were employed piecemeal. The extraordinary scenes of long columns halted for days on the highway north of Kyiv left Western observers bemused at the incompetence of Russian commanders. There was little attempt to disperse or move off-road, and the halted armour became easy meat for Ukrainian artillery strikes. Senior generals appearing close to the front to try and sort out the mess became casualties themselves, adding to the confusion. It seems apparent that poor tactics and training led to many of the losses, while a rigid 'top down' command structure stifled initiative in junior commanders.

Logistics

Adequate logistics is a vital component of modern warfare, and essentially it means getting the right ammunition, fuel and rations to the forward units to keep them fighting. Russian logistic support has proved largely inadequate in this conflict, adding to the poor morale of the frontline troops. It has become commonplace for tanks and armoured vehicles to be abandoned simply due to a lack of fuel or spares, and poor maintenance and chronic corruption in the supply chains have just added to the difficulties. When it comes to getting the supplies forward, the humble truck is a vital resource, but Russia's choice of cheap Chinese tyres has left many of them stranded by the roadside. The fact is large quantities of spare parts are required to keep modern mechanised forces moving, something that tends to be ignored in peacetime. In the First Gulf War, much of the Rhine army was stripped out of spares and engine packs to support the 1st Armoured Division deployed in the desert. Likewise, the Russians have had to cannibalise tanks held in storage for their vital spare parts. The production of new tanks has meanwhile been drastically hindered by the imposition of Western sanctions. As US General Patton famously remarked in World War Two, 'my men can eat their belts, but my tanks have gotta have gas'. This remains as true today as it was back in 1944, and poor logistical planning is one of the main contributors to the Russian failures in Ukraine.

Leadership

Despite the widely publicised modernisation of the Russian armed forces, the conflict has laid bare its continued reliance on old Soviet-style centralised command structures. Junior officers are taught blindly to follow orders, while there is no effective NCO corps in the sense of the British and US model. It appears that many of the soldiers taking part had no idea they were invading another country at the start of operations. These rigid command structures work fine until there is a reverse on the battlefield and plans start to go awry. This is why Western armies teach 'mission command', in which soldiers know the overall plan and can use their initiative to achieve the aim. This only works, of course, if you have well-trained junior leaders who are allowed to make command decisions, something totally lacking in the Russian system. Overlaying the leadership issues at the tactical level is a failure of strategic planning. The assumption seems to have been that the Ukrainian troops would collapse within a few days, with the government in Kyiv being rapidly decapitated. When this didn't happen due to the stubborn resistance of the Ukrainian Armed Forces, however, there was no Plan B. Western military observers have also noted that Russia has insufficient forces available to subdue a country the size of Ukraine. The multiple widely dispersed axis of advance worked against a concentration of force and served to further hamper the logistic effort. Trying to sort out the mess of the failed drive on Kyiv and other fronts led to the deaths of over 100 Russian generals and colonels, something unprecedented in modern conflicts. Perhaps the biggest failure in leadership came from the top, with Vladimir Putin himself. It appears that he truly believed he could achieve an easy victory and a divided West would simply stand by. However, he has been proven gravely mistaken, and NATO has found a new sense of purpose in opposing his expansionist policies.

Conclusion

The devastating armoured losses suffered by the Russians in Ukraine are not down to one decisive factor. Rather, they are the result of a combination of failures starting right at the top with Putin and his generals. The belief that the Ukrainians would quickly fold in the face of Russian aggression has been proved completely mistaken, and as a result, the flawed strategic plan behind the invasion quickly unravelled in the face of determined opposition. At the tactical level, the Russians failed to follow their own doctrine and left their armour fatally vulnerable. They were savagely punished for this by the Ukrainians and their Western-supplied anti-tank weapons. This was backed up by devastating artillery strikes directed by drones, which have made a significant contribution to the fighting. Meanwhile, a combination of failed logistics and poor morale has allowed significant quantities of Russian tanks and military material to fall into enemy hands. The Ukrainians have proved adept at reusing this gifted equipment and turning it against its former owners. The pre-war May Day Parades and video footage of massed armour gave the impression the Russian army was a powerful and modern fighting force. The harsh reality of the battlefield has shown it to be a largely un-reconstructed and poorly trained Soviet-style military. In fact, Western observers have been surprised at just how poorly it has performed, the opposite of pre-war predictions. The sight of so much destroyed and blasted armour led to the inevitable cry of 'the day of the tank is over'. This has proved wide of the mark, with both sides actively seeking to boost their tank numbers and the Russians even dragging veteran T-62s out of store. The decisive difference between the opposing sides is how these tanks are employed on the battlefield and in the end, and this is what has counted most.

A T-72 destroyed by a Western-supplied anti-tank missile, the fate of so many Russian tanks in Ukraine. *(Moscow Times)*

Other books you might like:

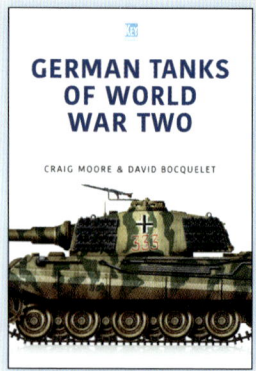
Military Vehicles and Artillery Series, Vol. 1

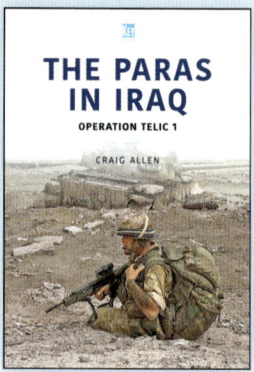
Modern Wars Series, Vol. 1

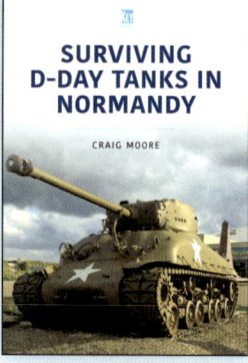
Military Vehicles and Artillery Series, Vol. 2

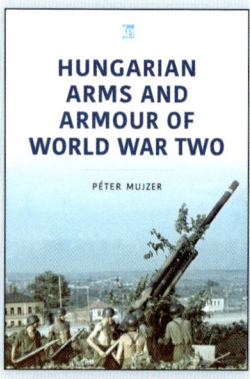
Military Vehicles and Artillery Series, Vol. 5

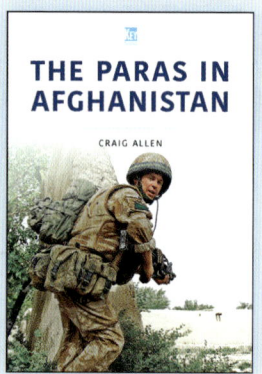
Modern Wars Series, Vol. 2

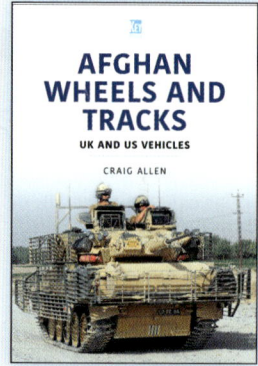
Military Vehicles and Artillery Series, Vol. 6

For our full range of titles please visit:
shop.keypublishing.com/books

VIP Book Club

Sign up today and receive
TWO FREE E-BOOKS

Be the first to find out about our forthcoming book releases and receive exclusive offers.

Register now at **keypublishing.com/vip-book-club**

Our VIP Book Club is a 100% spam-free zone, and we will never share your email with anyone else. You can read our full privacy policy at: privacy.keypublishing.com